Animals and Plants in Chinese Religions and Science

Huaiyu Chen

ANTHEM PRESS

Anthem Press
An imprint of Wimbledon Publishing Company
www.anthempress.com

This edition first published in UK and USA 2023
by ANTHEM PRESS
75–76 Blackfriars Road, London SE1 8HA, UK
or PO Box 9779, London SW19 7ZG, UK
and
244 Madison Ave #116, New York, NY 10016, USA

British Library Cataloguing-in-Publication Data
A catalogue record for this book is available from the British Library.

Library of Congress Cataloging-in-Publication Data
A catalog record for this book has been requested.
2022921105

ISBN-13: 978-1-83998-501-0 (Hbk)
ISBN-10: 1-83998-501-1 (Hbk)

Cover Credit: The Mural of West Wall, Yulin Cave 32.
Courtesy of *Dunhuang shiku quanji*, 2000.

This title is also available as an e-book.

CONTENTS

ACKNOWLEDGMENTS

Many colleagues and friends across the globe have helped me at the various stages of preparing, writing, revising, and publishing the materials in this book. I am profoundly grateful to my advisor Stephen F. Teiser for his wisdom and compassion in introducing me to the broad study of Chinese religions. Then my two colleagues at Arizona State University, Stephen B. Bokenkamp and Hoyt C. Tillman, have helped my research career in numerous ways. A distinguished visiting professorship from Henan University's Center for the Yellow River Civilization and Sustainable Development enabled me to spend three summers working on my project with some Chinese colleagues in China. I am grateful to Min Xiangpeng for facilitating my stays. Furthermore, I learned much about thinking with animals in Chinese religions from the conversations with Meir Shahar and Keith Knapp in the past several years. Finally, I would like to express my gratitude, especially to the following colleagues and friends who directly helped me with the chapters and this book: Annette Giesecke, Xudong Hou, Stephen N. Kory, Chien-hui Li, David Mabberley, Xinjiang Rong, Wutian Sha, Takata Tokio, and Alain Touwaide. The editorial team at the Anthem Press patiently guided me through the entire reviewing, editing, and publishing process, which I deeply appreciated.

Some earlier versions of the chapters in this book have appeared in various venues as follows: "Plant Science and Technology in Medieval China," in *A Cultural History of Plants in the Post-Classical Era*. Ed. by Alain Touwaide. Vol. 2, Ch. 4 (London: Bloomsbury, 2022), pp. 79–99, reproduced with permission from Bloomsbury Academic, an imprint of Bloomsbury Publishing Plc; "A Buddhist Classification of Plants and Animals in Early Tang China," *Journal of Asian History* 43: 1 (2009), pp. 31–51; "Animals, Divination, and Climate: An Environmental Note on the Cult of the Pig in Ancient China," *ISLE: Interdisciplinary Studies in Literature and Environment* (Oxford University Press, 2021), reproduced with permission from Oxford University Press; "The Changing Images of Zodiac Animals in Medieval Chinese Buddhist Literature," *Nesir Journal of Literary Studies* 1 (2021), 37–54; Review. *Unfortunate Destiny: Animals in the Indian Buddhist Imagination*, by Reiko Ohnuma, *Journal of*

Asian Studies 77: 3 (Cambridge University Press, 2018), pp. 828–30, reproduced with permission; Review. *Animals through Chinese History: Earliest Times to 1911*, ed. by Roel Sterckx, Martina Siebert, Dagmar Schäfer, *Journal of the Royal Asiatic Society*, 3rd series, 30: 2 (Cambridge University Press, 2020), pp. 376–78, reproduced with permission; Review. *Insects in Chinese Literature: A Study and Anthology*, by Wilt L. Idema, *China Review International* 26: 1 & 2 (University of Hawai'i Press, 2019), pp. 78–81; Review. *Panda Nation: The Construction and Conservation of China's Modern Icon*, by E. Elena Songster, *China Review International* 26: 1 & 2 (University of Hawai'i Press, 2019), pp. 110–14; Review. *Mao's Bestiary: Medicinal Animals and Modern China*, by Liz Chee, *Chinese Historical Review* 29: 1 (Routledge, 2022), pp. 67–69; and Review. *Healing with Poisons: Potent Medicines in Medieval China*, by Yan Liu, *China Review International* 27: 2 (University of Hawai'i Press, 2022), pp. 124–27. I thank these journals and publishers for their permission to reproduce some materials in this book.

My family is the strong pillar that supports my personal and intellectual life. My daughter Sarina is a talented artist. Many of her creative paintings inspired me to think about some topics on animals and plants. My wife Fei and my son Linden also brought me much joy, and I feel blessed by their love.

LIST OF ILLUSTRATIONS

Figures

Charts

INTRODUCTION

This book is a unique transdisciplinary study on animals and plants in medieval Chinese religions and science, especially in today's critical era of environmental crisis. In recent years, environment historians have written intensively on China. Yet, the study of animals and plants in medieval China was less developed, not much about the role of religions, more precisely. This book aims to bridge the gaps between religious studies and environmental studies, the history of science and religious studies, and animal studies and plant studies. It examines the biological, cultural, and spiritual encounters of animals and plants with humans in the medieval period through the analysis of changing roles and images of animals and plants in the historical, psychological, and imaginative experiences of human life, which are often overlooked in conventional scholarship.

This book can be viewed as a sibling of my recent book, *In the Land of Tigers and Snakes: Living with Animals in Medieval Chinese Religions* (New York: Columbia University Press, 2023), and its discussions move beyond animals to cover plants. This new book will look at how religious agents responded to the challenges of animals and plants as material culture in the mundane world. It will also look at how religious writers developed their various discourses about animals and plants from state ideology, and how the spiritual and natural worlds mutually enriched each other in China's medieval world. In particular, this book aims to analyze ordering nature and supernature in medieval China. This study has benefited tremendously from many historians of Chinese science and technology, such as Joseph Needham, Ho Peng Yoke, Nathan Sivin, and others on science and religions in China, and Berthold Laufer and Edward H. Schafer on plants and animals and material culture in medieval China along the Silk Road. In the meantime, this study attempts to raise new questions with the rise of animal and plant studies in contemporary scholarship. Many other readers in the domains of history of science and technology, Chinese history, Chinese literary culture, Chinese ideas and religions, animal studies, and material culture, in general, might find some interesting themes in my discussion. This book is centered on the following four interrelated topics: nature and supernature, the profane and sacred, religions and nonreligious thoughts, and Chinese tradition and the traditions from the Western Regions or "exotic territories (regions beyond the Chinese empire)."

For the first issue, I will focus on animals and plants and examine their representations and participation in medieval China's historical and supernatural life. Conventional scholarship often pays more attention to nonhuman life forms in nature in the disciplinary context of the history of science and technology but might have underestimated the significance of animals and plants in the psychological and imaginative experience in the context of medieval Chinese religions. Then, the further issue is that both animals and plants exited or created in the profane and sacred realms might equally enrich the material and spiritual life of the medieval world. My discussions in this book will reveal how medieval Chinese religious agents mobilized animals and plants to create and reshape the connections and transformations among humans, animals, and gods. Furthermore, religious and nonreligious communities mutually developed their textual and empirical knowledge of animals and plants. This knowledge and the communities produced, both religious and nonreligious, were often intermingled and enriched with each other. For example, medieval Chinese religious communities always attempted to come to terms with the Yin-Yang and Five-phase theories rooted in ancient Chinese cosmology. Lastly, since the focus of this book was the medieval world, the cultural and religious exchanges via the Silk Road were particularly vital in remapping animals and plants in China, with the introduction of new life forms from the neighboring regions, or "exotic territories."

In dealing with these four significant issues, this book is divided into three parts, each consisting of two chapters, plus a concluding chapter discussing recent trends in the field. The first part focuses on plants. The second part focuses on animal divination, and the last part focuses on the medieval transformation of animals in light of traditional Chinese cosmology. Chapter 1 argues that plant science and technology in medieval China cannot be separated from the developments in agriculture, culture, economy, and medicine since numerous species of plants were observed, cultivated, harvested, and used in the vast land of China. Besides indigenous plants, many from other parts of Asia were introduced into China and East Asia. This chapter highlights some crucial food plants, fabric plants, flowers, trees, and wood that played vital roles in medieval Chinese people's socioeconomic and daily life. From the perspective of the history of science and technology, this chapter explores various scientific and technological aspects of these plants, such as how medieval people knew about them, cultivated them, grew them, harvested them, and used them for socioeconomic purposes. These plants were also crucial in medieval Chinese people's political, cultural, and religious life.

Chapter 2 ventured into the medieval Chinese Buddhist monastic community for a case study on the plant knowledge of a Buddhist master Daoxuan (596–667). This chapter examines how Daoxuan as a medieval Buddhist master combined his Buddhist textual knowledge, his Chinese cultural background, and his empirical knowledge. This will be achieved from daily experience to develop a systematic classification of grains, vegetables, fruits, and other plants to maintain the monastic community's material foundation and spiritual integrity. My discussion reveals strong anthropocentrism in the medieval Chinese Buddhist community.

Chapter 3 moves to the animal divination in the socioeconomic life of ancient China. This chapter illustrates that the tradition of observing nature was deeply rooted in ancient Chinese cosmology centered on the Yin-Yang and the Five-phase theories. This cosmology was also linked to correlative theories that sought to establish resonant relationships between cosmic changes, environmental changes, social events, and human behavior. Hence, observing the behavior and health of animals was incorporated into the ancient arts of divination, including the technique of predicting weather changes. The observation of the close connection between animals and weather developed into worshiping animals. This chapter analyzes the rise of the pig cult, particularly concerning the context of environmental and climate changes in ancient China.

Chapter 4 provides an overview of zoomantic practice in medieval China. In the premodern Chinese bibliographical system, there were two types of textual records: official and nonofficial documents, each of which contained numerous zoomantic procedures. Government officials frequently generated official documents that served political governance goals. Official histories, annals, institutional documents, and Confucian classics were among these documents. The authorship or editorship of these documents was often explicit. Nonofficial documents included strange writings, tales, legends, and religious documents from Buddhism and Daoism, which were often not compiled under the sponsorship and support of the court or government. Some of them might have been collected by literati but lost original authorship. They did not serve political motivations and objectives, reflecting how people understood and interpreted correlative cosmology by observing animal behaviors at the local or non-bureaucratic level. Although many cases of animal divination appeared in the biographies of officials, zoomantic practices were more commonly described in the treatises on Five phases in official histories. It should be noted that, even within these treatises, no specific section focused exclusively on zoomantic phenomena. Instead, some cases were recorded because the compilers viewed them as extraordinary events within the traditional framework of correlative cosmology.

Chapter 5 shows how medieval Chinese Buddhist literature transformed twelve animals from compassionate deities into fallen angels to accommodate Buddhist ideas from South Asia for serving the Chinese audience. In an early Buddhist text, twelve animals were portrayed as the reincarnations of Bodhisattvas who traveled to teach and enlighten sentient beings who were reborn in the bodies of animals. Later on, medieval Chinese Buddhist commentators incorporated indigenous Chinese thoughts of Yin-Yang and Five-phase theories to reinterpret these animals as demonic spirits who could corrupt the morality of Buddhist practitioners. This new development manifested the attempt of medieval Chinese Buddhist monks to help Buddhist ideas take root in Chinese society.

The were-tiger legends in China and elsewhere are revisited in Chapter 6. The goal of this chapter is to rethink earlier work on were-tigers in medieval China from a two-dimensional perspective. First, it presents a novel interpretation of the were-tiger mythology as a cultural creation that incorporated ancient Chinese understanding of nature, human civilization, and the human body. Traditional Chinese Yin-Yang and Five-phase theories provided intellectual resources for interpreting the dynamic behind the were-tiger legend. Second, from a trans-Asian perspective, this study investigates the were-tiger legends in Korea, India, and Vietnam by comparing the shared narratives with the Chinese version and attempts to trace their historical connections.

Chapter 1

PLANT SCIENCE AND TECHNOLOGY IN MEDIEVAL CHINA

Introduction

Plant science and technology in medieval China cannot be separated from the developments in agriculture, economics, and medicine, as well as cultural practice. The Chinese empire ruled most of East Asia in the medieval period. Numerous species of plants were observed, cultivated, harvested, and used in the vast land of China that spanned a wide range of biomes from boreal to temperate and tropical, with most regions classified as subtropical. Besides indigenous plants, many plants from West, Central, South, and Southeast Asia were introduced into China and East Asia in general. This chapter focuses on particular food plants, fabric plants, flowers, trees, and wood which played vital roles in the socioeconomic life of medieval Chinese people. It highlights various scientific and technological aspects of these plants, such as how medieval people knew about them, cultivated them, grew them, harvested them, and used them for socioeconomic purposes. These plants were also important in the political, cultural, and religious life of medieval Chinese people. As covered in other chapters of this volume, some of these plants were introduced to Korea and Japan and became East Asian plants in general.

From the perspective of plant science and technology, there were several important changes and breakthroughs in medieval China, particularly between the sixth and fourteenth centuries. First, there was a spatial shift in terms of the main agricultural production areas from the north region centered on the Central Plain to the lower Yangtze Valley and Sichuan. In south China, rice-farming spread and multiple-cropping systems which yielded high rice production were used. While in north China, the more prolific and superior winter wheat which could be harvested three times in two years replaced millet to become the major food crop. These new developments made it possible to provide sufficient food supplies to the fast-growing population of the Chinese Empire; in the twelfth century, China surpassed a landmark when its population grew beyond 100 million. The development and expansion of agricultural production in south China might be due to several factors. Politically, there were three big waves of southward immigration that occurred: in the fifth century when the Western Jin 西晉 court lost control of the Central Plain and moved to Jiankang 建康 (modern Nanjing, Jiangsu Province); in the ninth century when the Tang Empire collapsed; and in the thirteenth century when the Mongols defeated the Southern Song Dynasty. In terms of technical innovation, as early as the sixth century, the Sui regime started building a canal system across the empire which became a major transportation route facilitating economic connection between the Central Plain and the lower Yangtze Valley. In the eighth century, after the An Lu-shan 安祿山 Rebellion, the Tang Empire was forced to shift its economic productivity to the southern regions, especially the lower Yangtze Valley and Sichuan areas. Agriculturally, in the tenth and eleventh centuries, more rice fields were developed in the Jiangnan area, especially the lower Yangtze Valley.[1]

Second, there was a long-term change in Chinese sericulture. Traditionally, the upper class valued silk textiles more than hemp and ramie fabric textiles. Hemp and silk were two of the most common fabric source materials. Mulberry (*Morus alba*), as the main source for providing leaves to feed silkworms, was a practical and symbolic plant for Chinese sericulture. Hemp (*Cannabis sativa*) was widely used for making oil, fabric cloth, and paper. Although silk production reached maturity in the Tang Dynasty and flourished, cotton was introduced to China in the eleventh century and its cultivation grew gradually. By the fourteenth century, cotton eventually became a significant fabric plant for the Chinese textile industry.

Third, with the development of agricultural production in south China, the cultivation and processing of tea (*Camellia sinensis*) began to increase. Tea culture appeared to become a significant part of daily life across the boundaries of various social classes. The established religious traditions such as Buddhism and Daoism played vital roles in this social

---

1 Peter J. Golas, "Rural China in the Song." *Journal of Asian Studies* 39, no. 2 (1980): 291–325; Robert M. Hartwell, "Demographic, Political, and Social Transformations of China, 750–1550." *Harvard Journal of Asiatic Studies* 42, no. 2 (1982): 365–442; William Guanglin Liu, *The Chinese Market Economy, 1000–1500* (Albany: State University of New York Press, 2015), 141–96; Xiongsheng Zeng, "Agriculture," in *A History of Chinese Science and Technology*, vol. 1, ed. Lu Yongxiang (Shanghai: Shanghai Jiaotong University Press, Berlin and Heidelberg: Springer Verlag, 2015), 351–429; Richard von Glahn, *The Economic History of China: From Antiquity to the Nineteenth Century* (Cambridge: Cambridge University Press, 2016), 208–54.

and cultural change. Both literati and clergies contributed to the rise of tea culture in the Tang and Song Dynasties. They discussed philosophies, religions, and literature while drinking tea. Numerous well-known poems on tea were composed. South China became the most critical region for cultivating tea trees. Various tea fields covered hills and mountains across the southern regions.

Fourth, in the Tang Dynasty, the expansion of the Chinese Empire and the development of its tribute system motivated gift exchange activities between the Tang court and other regimes as well as religious exchanges. These activities brought many exotic materials to Central China from abroad, including jewels and medicines, animals and plants, and science and technology. Fruit such as litchi (*Litchi chinensis*) and grapes (*Vitis vinifera*) and wood such as sandalwood (*Santalum* spp.) and rosewood (*Dalbergia* spp.) were introduced to China and became favorite luxuries of the upper class.

Fifth, expansion of religious monasticism and private gardens, religious buildings, temples, offices, houses, pavilions, bridges, boats, and other buildings occurred to a significant extent in the Tang and Song Dynasties not least because of the demands of the growing population. For instance, many medieval Buddhist monastic establishments received gardens and fields as well as animals and plants as donations; therefore, monks had to handle the classification and acceptance of these donations.[2] Many gardens and buildings increasingly used wood rather than stone for construction in the Tang and Song Dynasties. Wooden architectural technology became sophisticated in the Song Dynasty, which was mainly manifested by the appearance of a technical manual entitled *State Building Standards* (*Yingzao fashi* 營造法式) compiled by Li Jie 李誡 (1165–1110) in the twelfth century.

It should be noted that the expansion of agricultural cultivation toward the lower Yangtze River region and the west resulted in large-scale deforestation. Further, as already noted, wood was also extensively used as building architectural materials. Some of the earliest extant Buddhist temples and pagodas in north China are made of wood, such as the East Hall of the Buddha's Light Temple 佛光寺 in Mt. Wutai 五台山, Shanxi Province (857 CE), the Southern Chan Temple 南禪寺 (782 CE), and the Wooden Pagoda in Ying County, also in Shanxi, and the front façade of Mogao Cave 莫高窟 444 in Dunhuang 敦煌, Gansu Province (926 CE). Parts of these structures have been renovated and amended extensively over the years, but their main wooden structures and frameworks have endured for more

2 Huaiyu Chen, "A Buddhist Classification of Plants and Animals in Early Tang China," *Journal of Asian History* 43, no. 1 (2009): 31–51.

than a millennium.[3] The maturity of wooden architectural technology resulted in demands from both government and private projects for large amounts of timber, spurring the rapid deforestation in the Song and later periods. Mark Elvin lays out three phases of deforestation in Chinese history.[4] He notes that deforestation was due to various reasons: to make space for fields to cultivate crops and vegetables; to provide timber for constructing houses, other buildings, and transportation means including boats and carts; to make coffins and religious constructions; to burn logs for cooking, heating, and industrial processing; and so forth. The first phase took place in the north during the second half of the first millennium BCE when some measure of public exploitation could be detected. The second phase happened in the medieval period when the economic revolution began in the Yangtze River region where the cultivation of paddy rice was tremendously increased. It was during this time that severe wood shortages were first reported in the lower Yangtze River Valley and some parts of the western Chinese Empire. This crisis was characterized by bitter quarrels between local communities over the use of forested areas. The third phase started in the seventeenth century and continued to the modern period. Overall, natural disasters, military campaigns and wars, social and economic developments, the explosion of population, and the expansion of agricultural activities, as well as the extension of religious monastic establishments led to extensive deforestation.

Finally, in terms of knowledge of plant science, it was remarkable that during the medieval period, Chinese writers and scholars produced a large number of agricultural and biological works that recorded the cultivation of various domestic and imported plants, the socioeconomic functions of different parts of plants, and the technology for processing plants to produce food, oil, and fabric materials.

Knowledge of Plants

In medieval China, besides numerous works on particular flowers, fruits, and trees, the knowledge of plants can also be found in numerous agricultural works. One example is the *Compendium of Important Rules for Four Seasons* (*Sishi zuanyao* 四時纂要) by Han E 韓鄂 (c. late ninth and early tenth centuries), which contains details about numerous agricultural technologies as well

3 Nancy S. Steinhardt, "The Tang Architectural Icon and the Politics of Chinese Architectural History," *The Art Bulletin* 86, no. 2 (2004): 228–54.

4 Mark Elvin, *The Retreat of the Elephants: An Environmental History of China* (New Haven: Yale University Press, 2004), 84–85.

as a large amount of biological knowledge. Some ethnobiological works focused on geographical regions with exotic animals and plants based on the authors' own traveling experiences or local plants they observed and encountered. In the medieval period, many learned individuals traveled afar for the purposes of governing, exploring, or leisure. They left numerous records on local animals and plants. They expanded a genre of writing about exotic animals and plants, which was called the *Record of Exotic Things* (*yiwu zhi* 異物誌). For example, the *Record of the Land of Northward Facing Doors* (*Beihu lu* 北戶錄) by Duan Gonglu 段公路 (ca. late ninth century) and the *Record of Exotics in Lingbiao* (*Lingbiao luyi* 嶺表錄異) by Liu Xun 劉恂 (ca. late ninth century) both concern animals and plants in the Guangdong area. The former contains details about more than twenty plants, and the latter discusses more than ten plants and many types of fruit native to the Lingnan region. In the Song Dynasty, Fan Chengda 范成大 (1126–1193) wrote *Treatises of the Supervisor and Guardian of the Cinnamon Sea* (*Guihai yuheng zhi* 桂海虞衡誌, 1175) which describes plants in Guangdong and Guangxi.[5] Further, Song Qi 宋祁 (998–1061) wrote *Short Notes on Local Products from Chengdu* (*Yidu fangwu lueji* 益都方物略記) which contains details about plants in Sichuan. Li Gefei 李格非 (1045?–1106) prepared a book entitled *A Record of Celebrated Gardens in Luoyang* (*Luoyang mingyuan ji* 洛陽名園記) which discusses nineteen gardens in Luoyang and the numerous species of flowers they contained.[6]

More works on fruit and flowers appeared in the Song Dynasty. Some of the most prominent works are as follows: *A Record of Peonies in Luoyang* (*Luoyang mudan ji* 洛陽牡丹記) by Ouyang Xiu 歐陽修 (1007–1072 which focusses on *Paeonia suffruticosa* and was written in 1031 when he traveled to Luoyang; *A Register of Peonies of Tianpeng* (*Tianpeng mudan pu* 天彭牡丹譜) by Lu You 陸游 (1125–1210), composed in 1178 and containing details about thirty-four different peonies in Tianpeng, Sichuan; *A Register of Chinese Herbaceous Peonies* (*Shaoyao pu* 芍藥譜) by Liu Ban 劉攽 (1023–1089), who wrote about Chinese peonies (*Paeonia lactiflora*); *A Register of Plum (Meipu* 梅譜) by Fan Chengda (1126–1193) who discussed five categories of flowering plum (including *Prunus mume* "Alboplena," *P. mume* "Viridicalyx," and *P. mume* "Bungo"); *A Register of Chrysanthemum* (*Jupu* 菊譜) by Liu Meng 劉蒙

5 James M. Hargett trans., *Treatises of the Supervisor and Guardian of the Cinnamon Sea*, by Fan Chengda (Seattle: University of Washington Press, 2010).

6 Xiaoshan Yang, "Li Gefei's 'Luoyang mingyuan ji' (A Record of the Celebrated Gardens of Luoyang) Text and Context," *Monumenta Serica* 52 (2004): 221–55; Christian de Pee, "Wards of Words: Textual Geographies and Urban Space in Song-Dynasty Luoyang, 960–1127," *Journal of the Economic and Social History of the Orient* 52, no. 1 (2009): 85–116.

(ca. twelfth century) which was penned in 1104 and focused on 160 kinds of locally cultivated chrysanthemum flowers (*Chrysanthemum morifolium*); *A Register of Orchids from Jinzhang* (*Jinzhang lanpu* 金彰蘭譜) by Zhao Shiyu 趙時庚 (ca. thirteenth century), written in 1233 about 20 kinds of orchids; *A Register of Litchi* (*Lizhipu* 荔枝譜) by Cai Xiang 蔡襄 (1012–1067) in 1059 on 32 types of litchi from the Fujian area; *A Register of Citrus* (*Julu* 橘錄) by Han Yanzhi 韓彥直 (ca. twelfth century), penned in 1178 on 27 kinds of citrus (such as *Citrus x aurantium* Sour Orange Group, *C. reticulata* (mandarin), and *C. x aurantium* Sour Orange Group); and *A Register of Tong* (*Tongpu* 桐譜) by Chen Zhu 陳翥 (1009–1061) which was released in 1051 and was concerned with the cultivation and use of Paulownia (Paulowniaceae) trees. These works detailed traditional Chinese botanical classifications of flowers, fruit, and trees, rather than embodying any modern knowledge of plants. However, their meticulous descriptions of each kind or category (rather than the modern sense of species) of these plants can still help us understand and identify them by modern standards. More importantly, these books preserve extensive information about the cultivation and distribution of these plants across many regions of the Chinese Empire in the medieval period.

The knowledge of herbal plants has a long history in China. It became flourishing in the medieval period as the government started issuing the official versions of Materia Medica texts. The Tang court actively encouraged learning about traditional herbs. For example, in 659, the imperial court ordered Su Jing 蘇敬 (ca. seventh century) to organize a team to compile the *Newly Revised Materia Medica* (*Xinxiu bencao* 新脩本草). Literati who benefited from both traditional herbal learning and religious practice also extended their plant knowledge. Some of them devoted themselves to learning from the Materia Medica and left substantive written accounts. In the eighth century, Zhang Ding 張鼎 compiled one of the most interesting Materia Medica in the Tang Dynasty entitled *Materia Medica for Successful Dietary Therapy* (*Shiliao bencao* 食療本草), which appears to inherit a lot of information from the medical herbal learning of Sun Simiao 孫思邈 (581–682), a noted Daoist hermit and medical expert, and the prescriptions of Meng Shen 孟詵 (621–713), a medical expert and nutritionist who once served as a drafter for the emperor briefly. A fragment of this work was discovered in 1907 by Aurel Stein (1862–1943) from the cave library of Dunhuang, an oasis city in northwest China. It describes the medicinal and nutritional value of more than 160 food crops, oil, vegetables, and fruit, as well as other plants.[7]

7 Ute Engelhardt, "Dietetics in Tang China and the First Extant Works of Materia Dieterica," in *Innovation in Chinese Medicine*, ed. Elisabeth Hsu (Cambridge: Cambridge University Press, 2001), 184–87.

In the Song Dynasty, new pharmacopoeia with illustrations were compiled. Both medicinal and biological knowledge benefited from such endeavors. Further, traditional herbal texts continued to be written, annotated, compiled, and edited. Some writings on specific vegetables, fruit, and flowers appeared in the Tang and Song Dynasties, such as records of citrus, peony, bamboo, litchi, mushrooms, paulownias, and tea. Biological writings flourished in the Song Dynasty since many new works including herbs and trees or flowers and trees in their titles appeared, recording numerous plants in particular villas and private gardens across the country.[8] Since other chapters will cover medicinal plants, the main focus herein is on food plants, fabric plants, flowers, trees, and wood in the medieval period, particularly those that were more visible and remarkable than others.

Science and Technology of Food Plants

In the medieval period, Chinese agriculture witnessed the dominant rise of winter wheat (*Triticum aestivum*) and paddy rice (*Oryza sativa*) in the cultivation and production of grain crops. Before the Tang Dynasty, millet was the dominant traditional grain crop, which included two species: foxtail millet (*Setaria italica*) and broomcorn millet (*Panicum miliaceum*). Wheat gradually replaced millet and became the most important grain crop in north China after the mid-Tang period. Wheat was also cultivated in the southern regions. In the Yunnan area, there were even two harvesting seasons for wheat since the climate was warmer and more humid than in north China. In the moist regions of the Yangtze Valley, wheat was planted in the eighth month and harvested in the fourth or fifth month of the following year, same as in the north and the Central Plain. However, they could harvest three times in two years via a rotation system, which dramatically increased production. Paddy rice was also cultivated in north China. Still, the semiarid climate placed high demands on the artificial irrigation system, which impeded its popularity and sustainability, unlike its cultivation in south China.[9] In the Song Dynasty, even in the Yangtze Valley, technological innovation meant that wheat could be harvested twice per annum. In addition, both wheat and rice were cultivated

8 Liu Changzhi, "Zhongguo chuantong zhixue de fazhan (The Development of Chinese Traditional Biology)," in *Zhongguo zhiwu xue shi (A History of Plant Science in China)*, ed. Zhongguo zhixue xuehui (Chinese Association for Plant Science) (Beijing: Kexue chubanshe, 1994), 35–59; Guihuan Luo, "Biology," in *A History of Chinese Science and Technology*, vol. 1, edited by Yongxiang Lu (Heidelberg: Springer, 2015), 446–53.

9 Ling Zhang, *The River, The Plain, and the State: An Environmental Drama in Northern Song China* (Cambridge: Cambridge University Press, 2016), 233–39.

respectively in different regions and seasons in the south, which enhanced the ability to support the rapidly growing population of the Chinese Empire.

In the Tang Dynasty, two issues seem to have contributed to the development of wheat and rice cultivation in the south. First, large-scale immigration from north China brought abundant agricultural labor to the south. Numerous fields were developed and expanded. Second, in the Sui and Tang Dynasties, the cultivation of wheat and rice first developed along with the canal system that linked the Central Plain and the lower Yangtze Valley. The skilled immigrants settled in the lower Yangtze Valley propelled the cultivation of these grain crops. According to the *Old History of the Tang Dynasty* (*Jiu Tang shu* 舊唐書), in the mid-Tang period, ten times more rice was transported from the south via the grand canal than in the early Tang era. Farmers in the south also developed more diverse rice fields to ease demands on irrigation. For example, the technology of the ladder fields (*titian* 梯田) probably first appeared in Yunnan in the Tang period and then flourished across all of south China in the Song Dynasty.[10] Technically, the expansion and improvement of rice cultivation in the Song Dynasty was due to many factors such as water control, farm implements, draft animals, planting methods, weed and pest control, manuring, seed selection, and human innovation.

Paddy rice became the principal grain crop in the Song Dynasty. It should be noted that the introduction of Champa rice from Southeast Asia into the Fujian area in the eleventh century became a game changer. This rice can be double-cropped in the southern regions of China. Because of its early ripening, it also matured much faster than traditional paddy rice. Champa rice offered better drought resistance qualities, enabling it to grow in deficient soil. Its flexibility to cope with flooding and salinity also helped it tolerate more abiotic stresses for surviving, maturing, and harvesting. Therefore, its cultivation was swiftly extended to the lower Yangtze Valley.[11] In the twelfth century, paddy rice eventually took the leading role across the country. The increasing production of both winter wheat and paddy rice laid the foundation for supporting a large population in the twelfth century when the people of Song China probably just passed the landmark of 100 million.

In the medieval period, soybean (*Glycine max*) was another important food plant, which often served as a supplement to winter wheat and paddy rice. However, it was also important for oil production. The technology for extracting oil from soybeans was already established in medieval China and there were

10 Francesca Bray, *Science and Civilisation in China, Volume 6: Biology and Biological Technology, Part II: Agriculture* (Cambridge: Cambridge University Press, 1984), 126.

11 Randolph Barker, "The Origin and Spread of Early-Ripening Champa Rice: Its Impact on Song Dynasty China," *Rice* 4 (2011): 184–86.

two major plants for extracting oil in daily life which were widely cultivated across the country: Sesame (*Sesamum indicum*) and choy sum (*Brassica rapa* Chinensis Group, Chinese name: youcai 油菜). Sesame was the principal oil plant until the thirteenth century when choy sum became the chief choice for farmers to plant as the winter vegetable. Sesame oil was widely used for daily cooking and medical healing in the medieval period. It seems that, although choy sum was first cultivated in the northwestern region for its resistance to cold winters (Chinese farmers also called it "hancai 寒菜," cold vegetable), it was introduced into the Yangtze Valley and was expanded in the southern regions in the Song Dynasty. Its leaves were used as food, but its seeds were pressed for making oil. Farmers used the oil for lighting lamps and also washing hair, and in this respect it seems to have been a better option than sesame oil. From the thirteenth century, it became the dominant oil plant.

The rise of tea culture in the Tang and Song Dynasties marked one of the most important developments in the history of Chinese plants, economically, socially, culturally, and diplomatically. The cultivation and production of tea trees originated in southwest China, especially Yunnan and Sichuan. In early medieval China, Daoist clergy started consuming tea as a drink made from young leaves of tea trees. Aristocratic literati accepted this beverage and introduced it to the Chinese court. Even aristocratic women enjoyed it. Tea then became a typical drink for the upper class in the early medieval period. In most cases, tea was made from the leaves of wild tea trees that grew naturally in the mountains. Later in the Tang Dynasty, most tea producers grew domesticated tea trees. Therefore, the cultivation and production of tea flourished across the country, especially in the Yangtze region.

In the Tang Dynasty, the technology of cultivating tea trees was also improved. Around 760–762, Lu Yu 陸羽 (733–804) wrote a book entitled *The Classic of Tea* (*Chajing* 茶經) offering a detailed account of the cultivation of tea trees and the production of tea beverages as well as the fashion of tea drinking as a culture. His horticultural description touched on the climatic, soil, and water conditions for cultivating tea trees. His book also discussed the technology of transplanting tea trees and harvesting tea leaves. Lu Yu listed fifteen tools for picking, steaming, pressing, drying, and storing tea leaves and cakes and the methods of making tea cakes. This allows us to learn many specific details about how tea beverages were made in the eighth century.[12]

12 Victor H. Mair and Erling Hoh, *The True History of Tea* (New York and London: Thames & Hudson, 2009); James A. Benn, *Tea in China: A Religious and Cultural History* (Honolulu: University of Hawai'i Press, 2015), 21–41.

In the ninth and tenth centuries, the technology of cultivating tea trees and water irrigation became more mature and sophisticated, and tea cultivators even started harvesting tea leaves in the tenth month of the lunar calendar year, besides the second, third, and fourth months. Tea production increased to meet the demand of a large population of consumers. Eight percent of the local prefectures in the Tang Dynasty cultivated tea trees and produced tea in the eighth century. In the Song Dynasty, more farmers were involved in growing and producing tea and the government instituted tea households as a new category in its household registration system.

In the Tang and Song Dynasties, the Yangtze and Southeast regions such as Zhejiang, Fujian, Jiangxi, Hunan, Hubei, Anhui, and Sichuan were the main places for cultivating and producing tea. In the ninth century, there were probably four kinds of gardens for growing and producing tea: government tea gardens, private tea gardens of landlords, tea gardens of Buddhist monasteries, and self-supported tea gardens of farmers. The quality of these tea products varied depending on the region, technology, and the skill of the cultivator. The popularity and success of tea resulted in the government imposing a tax on tea production and trade.[13] The rise of tea culture certainly laid the foundation for the status of tea as an iconic product of the Chinese Empire in later periods.

Plants for Seasoning and Aromatics

As Edward H. Schafer noted, there was little clear-cut distinction among drugs, spices, perfumes, and incenses, because all of them were used to nourish both body and spirit.[14] Many plants were used as sources for making seasoning and aromatics. Medieval Chinese people enjoyed cooking their food using various types of spice and seasoning, which came from a broad array of aromatic plants such as ginger (*Zingiber officinale*), green onion (*Allium chinense*), Sichuan pepper (*Zanthoxylum simulans*), garlic (*Allium sativum*), Chinese parsley (*Coriandrum sativum*), garlic chives (*Allium tuberosum*), and so on.[15] *The Monthly Ordinances for Four Classes of People* offers detailed descriptions of the stages and skills in growing these vegetables, including the selection and planting of seeds, transplanting, irrigation, and fertilizer use. In the thirteenth century,

---

13 Wang Hongjun, "Tangdai de chaye shengchan (The Tea Production in the Tang Dynasty)," *Qilu xuekan* (*Journal of Qi Lu*), no. 6 (1987): 14–21.

14 Edward H. Schafer, *The Golden Peaches of Samarkand: A Study of T'ang Exotics* (Berkeley: University of California Press, 1963), 155.

15 Li Fan, *Zhongguo zaipei zhiwu fazhan shi (A History of the Development of Cultivated Plants in China)* (Beijing: Kexue chubanshe, 1984), 130–40.

the technology of cultivating chives in greenhouses was improved by introducing horse manure to fertilize soil.

In the medieval period, aromatics were widely used among royal families, bureaucrats, religious adherents, businesspeople, and the military. Different types of incenses commonly appeared in political, military, and religious rituals, especially sacrificial and funeral ceremonies. According to Schafer, among numerous native plants and animals that the Chinese used for producing perfume and incense, cassia (*Cinnamomum cassia*, also called Chinese cinnamon), camphor (*Cinnamomum camphora*), and liquidambar (*Liquidambar acalycina*) were commonly used.[16] However, other aromatic plants were introduced to medieval China from abroad as exotic indigenous tribute. One of the most popular incenses was imported agarwood (*Aquilaria* spp.) from southeast Asia.[17] Both Buddhist and lay communities highly valued this type of fragrant wood in their daily use of incense. Aristocratic people even developed a fashion for wearing a bag containing this incense. Agarwood was used for making statues of Buddha saints for its fragrance, as well as highchairs for eminent monks to sit on while they preached Buddhist teachings. Clove (*Syzygium aromaticum*) from Indonesia was used as both a culinary spice and an incense in the Tang Dynasty. In the underground chamber of Dharma-Gate Temple (Famensi 法門寺) near Xi'an, Shaanxi Province, archaeologists discovered that clove and incenses were offered. Other exotic aromatic plants included lakawood (*Dalbergia parviflora*) from Indonesia, elemis (*Canarium* spp.) from southeast Asia, storax (*Storax balsam*) from Asia Minor, benzoin (*Styrax*) from Indochina and Indonesia, frankincense (*Boswellia* spp.) from Arabia, and putchuk (*Saussurea costus*) from Sri Lanka. Nevertheless, these perfumes and incenses might not have been known to all social classes, but they were favored by the upper classes.

Textile Plants

In traditional China, many plants, including hemp, ramie (*Boehmeria nivea*), chingma (*Abutilon theophrasti*), jute (*Corchorus capsularis*), and flax (*Linum usitatissimum*) were used as fabric materials. It is beyond doubt that hemp was the most commonly used plant for cloth-making and papermaking in medieval China. As Joseph Needham noted, the "hemp" or the Chinese "ma 麻" nexus might include more than twenty plants.[18] Although

16 Schafer, *The Golden Peaches of Samarkand*, 158.

17 Schafer, *The Golden Peaches of Samarkand*, 163–65.

18 Joseph Needham, *Science and Civilisation in China, Volume 6: Biology and Biological Technology, Part I: Botany*. With the Collaboration of Lu Gwei-Djen and a special contribution by Huang Hsing-tsung (Cambridge: Cambridge University Press, 1986), 170–74.

the identification of hemp in ancient Chinese sources was often a puzzle due to the ambiguous meaning of the Chinese character "ma," scholars have combined textual sources and extant species from manuscripts found in Dunhuang for tracking it down. In medieval China, the flax was often called "barbarian flax (huma 胡麻)" in Chinese sources. There was both cultivated hemp and wild hemp in China.

The cultivation of hemp has a very long history as it was a popular plant in the Shang and Zhou Dynasties (eleventh to third centuries, BCE). At that time, it was commonly cultivated in the middle and lower regions of the Yellow River. But around the common era, its cultivation was expanded to the southern regions of the Yangtze River. In the medieval period, hemp was cultivated throughout the entire Chinese Empire. It continued to be cultivated in other frontier regions until the thirteenth century when the central government began to encourage the cultivation of cotton. Some early agricultural sources including the *Book of Si Sheng* (*Sisheng zhishu* 氾勝之書) and *Essential Techniques for the Common People* (*Qimin yaoshu* 齊民要術, ca. 535) noted methods for cultivating hemp to ensure hard stalk, thick bast, and more internodes in order to harvest strong hemp raw materials for textile production; indeed, these sources offer many processual details covering seed selection, seed soaking, intercropping, sowing, and harvesting. The hemp seed and oil occasionally served as food and medicine for some families in north China.[19] According to some Dunhuang manuscripts dated in the ninth and tenth centuries, hempseed was used for oil pressing. Although hemp oil was used for both cooking and lighting, the quality was poor.[20]

Hemp was the main source for making cloth and paper in the medieval period. In early medieval China, the fiber made from hemp even functioned as money for paying taxes to the state government in northwestern and southern China.[21] Chinese families, mostly women, converted the hemp fiber into a textile through spinning and weaving. Knitting and felting might

19 Li Huilin, "The Origin and Use of Cannabis in Eastern Asia Linguistic-Cultural Implications," *Economic Botany* 28, no. 3 (1974): 293–301; Li Huilin, "An Archaeological and Historical Account of Cannabis in China," *Economic Botany* 28, no. 4 (1974): 437–48; Liu Fei-hu, Hua-ren Hu, Gang-hui Du, Gang Deng, and Yang Yang, "Ethnobotanical Research on Origin, Cultivation, Distribution and Utilization of Hemp (*Cannabis sativa* L.) in China," *Indian Journal of Traditional Knowledge* 16, no. 2 (2017): 235–42.

20 Eric Trombert, "Cooking, Dyeing, and Worship: The Uses of Safflower in Medieval China as Reflected in Dunhuang Documents," *Asia Major* 3rd series, 17, no. 1 (2004): 62.

21 Wang Binghua and Helen Wang, "A Study of the Tang Dynasty Tax Textiles (Yongdiao Bu) from Turfan," *Journal of the Royal Asiatic Society* 3rd series 23, no. 2 (2013): 263–80.

also be used for making textiles. Traditional spinning and weaving technology first involved warp and weft threads bound together with cloth beams made on a loom. There were usually three basic structures to this cloth, namely tabby (*pingwen* 平紋), twill (*xiewen* 斜紋), and stain (*duanwen* 緞紋).[22] Even as early as the Warring States period (475–221 BCE), both silk and hemp cloth were accepted as tax payments. This tradition continued, with various changes experienced during the early medieval period. In 737, in compiling a new code, the Tang government regulated hemp cloth as an alternative tax payment if silk cloth was not an option. An adult was required to pay 77.7-cm long hemp cloth and three pounds of raw hemp fiber per annum.[23]

Hemp was the most important raw material for making cloth in traditional China. Many scholars have noted that hemp remained as the generic name for plant fibers before cotton was introduced from Central Asia into China around the tenth to eleventh centuries. In the Tang Dynasty (618–907), four main types of establishments worked on textile production: rural households, official workshops, private urban workshops, and weaving households. In the eighth century, they contributed about 7.4 million bolts of tabby and more than 1.85 million hanks of silk floss to the state. As a typical labor division in medieval China, male peasants focused on cultivating hemp, ramie, and mulberries in the field, and female household members played the most important role in yarning and weaving for producing textiles. According to Hong Mai 洪邁 (1123–1202), when a young lady died at 21, she was buried with 33 bolts of open weave, 70 bolts of plain silk, and about 50 meters (about 164 feet) of coarse silk. The inventory demonstrated her production in her lifetime. In addition, it should be noted that the ancient Chinese cultural tradition required family members to dress in raw tabby clothing while mourning the death of a relative.[24]

Hemp was also a common raw material used for papermaking and printing technology in medieval China. Traditional Chinese historiography regards papermaking in proud terms since its history can be traced back to the inventor Cai Lun 蔡倫 (48–121). Although many textual sources have

22 Chen Weiji, *Zhongguo fangzhi kexue jishu shi (A History of Textile Production Technology in China)* (Beijing: Kexue chubanshe, 1984), 41–43.

23 Angela Sheng, "Determining the Value of Textiles in the Tang Dynasty: In Memory of Professor Denis Twitchett (1925–2006)," *Journal of the Royal Asiatic Society* 3rd series 23, no. 2 (2013): 182–84.

24 Liu et al., "Ethnobotanical Research on Origin, Cultivation, Distribution and Utilization of Hemp (*Cannabis sativa* L.) in China," 239–40; Sheng, "Determining the Value of Textiles in the Tang Dynasty," 182–84; Francesca Bray, *Technology and Gender: Fabrics of Power in Late Imperial China* (Berkeley: University of California Press, 1997), 189–200.

described the process of papermaking technology in ancient China, early paper products have not survived to the present day. However, various manuscripts written in multiple languages have been found by Aurel Stein, Paul Pelliot (1878–1945), and other Western explorers in the early twentieth century in the extraordinary cave library in Dunhuang and other sites of ruined cities in Chinese Central Asia. Many of these manuscripts are made from hemp, which was called small hemp paper for it might have been produced locally and was different from the paper used by the central government of the Tang Empire. Even though the manuscripts were found in Dunhuang, numerous documents were originally issued by the central government, and some religious manuscripts were distributed from the capital city of the Tang Empire, Chang'an. The paper from the central part of the Tang Empire, especially Chang'an, was better made with higher quality than what might have been made locally in Dunhuang.

Official documents issued by the central court of the Tang Empire were prepared on paper that was called "huangbo 黃檗" hemp paper; this was more durable than regular hemp paper. The "huangbo" paper was produced from hemp paper that was completely soaked in oil extracted from the Amur cork tree (*Phellodendron amurense*).[25] Along the Silk Road, many manuscripts used paper that was locally made. As Agnieszka Helman-Ważny's fiber analysis shows based on samples from 130 manuscripts, there were four distinct groups of rag paper which came from textile remnants.[26] Among these 130 manuscripts, only four were made using pure hemp. The majority of these manuscripts, 84, were made from ramie, some with a hemp addition. In the third group, there were eight cases where manuscripts were made of ramie and hemp with the addition of paper mulberry (*Broussonetia papyrifera* or a *Morus* sp.). The addition of paper mulberry fibers might have been due to a shortage of ramie and hemp in the northwestern region. The fourth group includes fifteen samples that were made of ramie and hemp with the addition of other, singular, fibers such as flax, jute, silk, or cotton. Clearly, hemp was the most important raw material for producing paper across the Tang Empire, with the addition of other fibers varying region by region depending on the supplies of available plant materials.

---

25 Jean-Pierre Drège, "L'analyse fibreuse des papiers et la datation des manuscrits de Dunhuang," *Journal Asiatique* 274 (1986): 403–15; Rong Xinjiang, *Eighteen Lectures on Dunhuang*, trans. Imre Galambos (Leiden: Brill, 2013), 485.

26 Agnieszka Helman-Wazny, "More than Meets the Eye: Fibre and Paper Analysis of the Chinese Manuscripts from the Silk Roads," *STAR: Science & Technology of Archaeological Research* 2, no. 2 (2016): 127–40.

Flax was another important fabric plant in medieval China. It was called "barbarian hemp" in traditional Chinese literature, which indicates that it was an exported plant from foreign lands. According to Chinese sources such as the *History of the Han Dynasty* (*Hanshu* 漢書), it was first brought to China by Zhang Qian 張騫 from a mission to Central Asia. However, some modern scholars question the veracity of this claim.[27] Instead, they suggest that flax was first cultivated as a fiber crop and then as an oil crop in India and China. The third-century agricultural work entitled *The Monthly Ordinances for Four Classes of People* (*Simin yueling* 四民月令) offers information about planting flax. In medieval China, flax became one of the most popular plants along the Hexi Corridor, the pathway region linking the central part of the Chinese Empire with its northwestern frontier regions. It was used by local people for making oil, paper, and ropes.

Two species of *cotton* were introduced into medieval China through different routes. One was the African-Asian species (*Gossypium herbaceum*) that came to China mainly through the Silk Road from Central Asia. Stephen Dale even suggested that the Silk Road should also be regarded as the Cotton Road.[28] In early medieval Chinese sources such as the social and economic documents among the manuscripts uncovered from Chinese Central Asia, cotton was called "*baidie* 白疊." In 1959, Chinese archaeologists discovered two fragments of cotton cloth from a second-century tomb near Niya 尼雅 in Xinjiang. As early as the sixth century, a local regime centered on an oasis city, Gaochang 高昌 Kingdom (modern Turfan), was known to produce cotton cloth. In the ninth and tenth centuries, this area often submitted cotton cloth as a tribute to the central court of the Chinese Empire. However, cultivation of this species never became widespread due to limitations of both environmental conditions and political control. The second species was the South Asian species (*Gossypium arboreum*), which came to southern regions of the Chinese Empire from India (especially from modern Assam and Eastern Bengal states) via Myanmar. After the fourteenth century, this species flourished across Southern China. In the medieval period, mostly in the Tang and Song Dynasties, cotton was not as popular and essential as in modern times. On the one hand, it was not an indigenous plant to the Chinese land. On the other hand, when some Chinese residents in the Tang Dynasty started to learn about cotton and its production, the use

---

27 Liu Fei-hu, Xia Chen, Bo Lng, Rui-yan Shuai, and Chun-lin Long, "Historical and Botanical Evidence of Distribution, Cultivation and Utilization of *Linum usitatissimum* L. (flax) in China," *Vegetation History and Archaeobotany* 20, no. 6 (2011): 561–66.

28 Stephen F. Dale, "Silk Road, Cotton Road or.... Indo-Chinese Trade in Pre-European Times," *Modern Asian Studies* 43, no. 1 (2009): 79–88.

of cotton garments was regarded by the government as a threat to preserving social status markers, since luxury silk and common hemp already occupied dominant positions to serve the main sources for making different garments to distinguish between the upper and lower classes. Moreover, official financial administrators also viewed cotton as a potential complicating hindrance to the fiscal regime. In the Song Dynasty, the cultivation of South Asian species of cotton was growing fast, especially in the lower Yangtze River region. In 1273, the government even issued a significant agriculture handbook entitled *Fundamentals of Agriculture and Sericulture* (*Nongsang jiyao* 農桑輯要) detailing the cultivation of cotton. In the thirteenth century, cotton cultivation and cotton textile production were encouraged by the government. Cotton cloth was adopted as an important material for making both military uniforms and civilian garments.[29]

Arguably, silk was the most reputable fabric material in medieval China, especially before the fourteenth century when cotton took over the dominant position in the textile production of China. Although silk came from silkworms and was thus not a plant, the leaves of a plant were used as the indispensable food for feeding silkworms. This plant was the white mulberry tree (*Morus alba*), one of the most extensively cultivated plants across the Chinese Empire for 4,000 years. The Chinese were devoted to developing its sericulture by cultivating mulberry trees, raising silkworms, weaving silk cloth, and producing silk textiles. Silk cloth functioned not only for making textiles but also for serving as money and taxes along the Silk Road, and it became significant financial sources of the Chinese government. According to the *Tang Code* that was made law in 737, in silk-producing areas, the per annum tax for a household was about 62.2 centimeters (about 24.5 inches) silk plus 85 grams (about 3 ounces) of silk floss. For those areas that did not produce silk, the tax was about 77.7 centimeters (about 30.6 inches) of hemp cloth plus 1.3 kilograms (about 2.86 pounds) of hemp fiber. It appears that in northern China, silk was more favored than coins and hemp cloth, yet in southern China, coins and hemp were more common as currency, but silk was more valuable.[30] Expensive silk was often dyed by using plants such as safflower (*Carthamus tinctorius*), indigo (*Indigofera tinctoria*), and purple gromwell (*Lithospermum erythrorhizon*).

Chinese sources often named two characters "nong 農" (agriculture) and "sang 桑" (mulberry) together and used the phrase "sangtian 桑田" (mulberry

---

29 Harriet T. Zurndorfer, "Cotton Textile Manufacture and Marketing in Late Imperial China and the 'Great Divergence'," *Journal of the Economic and Social History of the Orient* 54, no. 5 (2011), 705–6.

30 Sheng, "Determining the Value of Textiles in the Tang Dynasty," 175–95.

fields) for referring to cultivated lands, which highlighted the preeminent role of mulberry cultivation in traditional Chinese agricultural civilization. The central court required local officials to take action to encourage farmers to cultivate mulberry trees and household women to weave silk and produce silk textiles. For instance, in the thirteenth century, each male adult of a household was required to plant twenty mulberry (or) jujube (*Ziziphus* sp.) trees each year.[31] However, occasionally, in the medieval period, due to a shortage of mulberry leaves, silkworms could be fed with the leaves of oak trees which resulted in the production of coarser silk.

Besides serving as feedstock for silkworms, it should be noted that mulberry leaves were also used to feed cattle as well as being used as a convenient herbal medicine for healing wounds in medieval China. Mulberry trees also served as raw material for producing paper along the Silk Road.[32]

Cultivation of Flowers

Flowers were cultivated and used for many purposes in the medieval period. Traditional Chinese sources often mention ornamenting and viewing activities centered on the flower. Both textual and visual sources manifest the vibrant life of flowers among all social classes. Besides botanical records on flowers, medieval literati also wrote prose, poems, and lyrics about flowers they cultivated, observed, and discussed in their daily life. Visual materials such as murals in Buddhist caves and wall paintings in tombs also depict flowers. Some flowers seem to be more visible than others. For example, peonies were famous for their popularity in both large gardens and private villas in the medieval period.[33] Other flowers of fruit often appeared in poems too, such as those of the pomegranate (*Punica granatum*).[34]

Li Deyu 李德裕 (787–850), a high-ranking official in the late Tang Dynasty, developed his own private villa in Pingquan 平泉, about 10 miles south of Luoyang, modern Henan Province. Originally, he acquired a garden estate for commemorating his father for his filial piety, but soon later, he became obsessed with enjoying his reclusive residence surrounded by flowers and mountains. During his long-term career in the civil service and transferring

31 Dieter Kuhn, *Science and Civilisation in China, Volume 5. Chemistry and Chemical Technology, Part 9. Textile Technology: Spinning and Reeling* (Cambridge: Cambridge University Press, 1988), 286–88.

32 Helman-Wazny, "More than Meets the Eye," 132.

33 Yang, "Li Gefei's 'Luoyang mingyuan ji'," 221–55; de Pee, "Wards of Words," 85–116.

34 Donald Harper, "Flowers in T'ang Poetry: Pomegranate, Sea Pomegranate, and Mountain Pomegranate," *Journal of the American Oriental Society* 106, no. 1 (1986): 139–53.

from place to place across the Tang Empire, he collected numerous plants from locations where he was stationed and cultivated them in his Pingquan Villa.[35] Not all of these plants survived. However, textual sources indicate that at least sixty-nine plant species did survive there, among which modern scholarship has identified about fifty-nine, including both trees and flowers that vary from tropical plants (39%) to temperate plants (3%). Based on the descriptions from numerous poems by Li, some flowers can be identified as follows: dwarf sedge (*Acorus gramineus*), Tartarian aster (*Aster tataricus*), buddleja (*Buddleja officinalis*), camellia (*Camellia japonica*), Japanese glorybower (*Clerodendrum japonicum*), eupatorium flower (*Eupatorium fortunei*, Chinese name: *peilan*), Korean sweetheart tree (*Euscaphis japonica*), hibiscus (*Hibiscus mutabillis*), rose of Sharon (*H. syriacus*), coastal rose of Sharon (*H. hamabo*), big leaf hydrangea (*Hydrangea macrophylla*), Chinese green flower lily (*Lilium fargesii*), the flower of Kaido crab apple (*Malus x micromalus*), the flower of Magnolia (*Manglietia insignis*, *Magnolia biondii*, and *M. pilocarpa*), white lotus (*Nelumbo nucifera*), oleander flower (*Nerium oleander* [*N. indicum*]), ormosia (*Ormosia semicastrata*), sweet olive (*Osmanthus fragrans*), photinia (*Photinia serratifolia*), plum (*P. mume*), the flower of David's peach or Chinese wild peach (*P. davidiana*), Nanking cherry (*P. tomentosa*), the flower of Chinese white pear (*Pyrus bretschneideri*), rhododendron (*Rhododendron mariesii*), cloud-pattern silk rhododendron (*R. fortunei*), multiflora rose (*Rosa multiflora*), broadleaf lilac (*Syringa oblata*), sapphire berry (*Symplocos paniculata*), coltsfoot (*Tussilago farfara*), sea bilberry (*Vaccinium bracteatum*), the Chinese wisteria (*Wisteria sinensis*).[36] This long list illustrates the wealth of flowers that Li cultivated in ninth-century China. Many of them served as ornamentals for displaying and viewing, but some could produce fruit or herbs for medical use.

Among numerous flowers that were cultivated in medieval China, safflower occupied a prominent position for its use of multiple purposes. The discovery of numerous manuscripts from Dunhuang helps us understand that *safflower* played a sophisticated role in medieval Chinese economic, social, and cultural life. In 1993, Chinese scholar Wang Kexiao 王克孝 first identified it in a careful reading of some manuscripts. Then Eric Trombert offered a comprehensive study of this flower.[37] Together they demonstrate that safflower was used to

35 Xiaoshan Yang, "Li Deyu's Pingquan Villa: Forming an Emblem from the Tang and to the Song," *Asia Major* 3rd series, 17, no. 2 (2004): 45–88.

36 Feng Guangping, "Tangdai Pingquan shanju zaipei zhiwu kao (Textual Research on Cultivated Plants in Pingquan Villa of Tang Dynasty)," *Zhongguo lishi dili luncong (Journal of Chinese Historical Geography)* 31, no. 1 (2016): 60–72.

37 Eric Trombert, "Des fleurs rouges en galette: Une plante tinctoriale dans la Chine ancienne: le carthame." *Journal Asiatique* 285, no. 2 (1997): 509–47; Trombert, "Cooking, Dyeing, and Worship," 59–72.

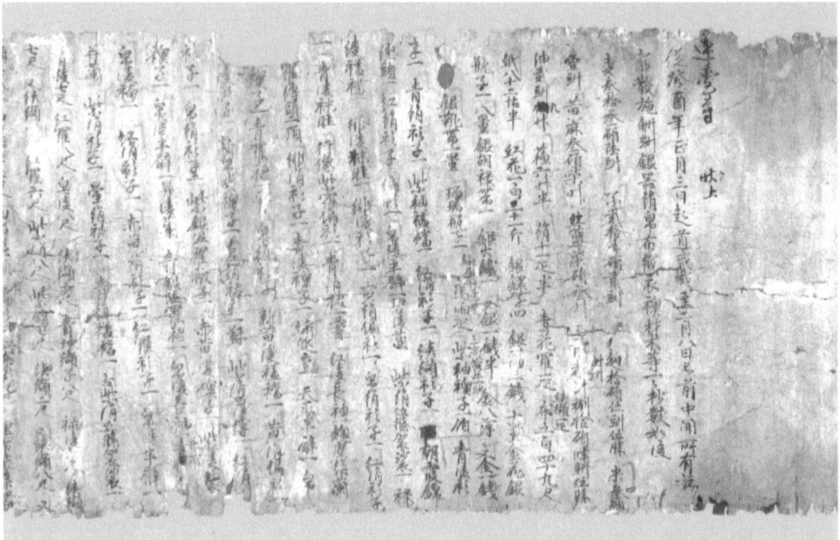

Figure 1.1: A monastic account record on the Dunhuang manuscript Pelliot chinois 2567v mentioned safflower. © The Bibliothèque nationale de France.

serve as a medical herb, to dye various textiles, to produce rouge (yanzhi 胭脂), and to extract oil for cooking purposes. Egyptians first cultivated safflower, followed later by the Persians. Eventually, it reached China via the Silk Road and flourished in northwestern regions (Figure 1.1). As early as in the *Essential Techniques for the Common People* in the sixth century, it was stated that oil could be extracted from safflower seeds to make candles and a lubricant for cartwheels and candles. It also suggested that once the safflower was harvested in the summer, it should be processed immediately.

In the ninth and tenth centuries, the use of safflower seeds and petals expanded. Besides serving medical and cooking purposes, safflower was also used for making dyestuff. According to Trombert's study on a letter in the early tenth century, rouge was listed as the most expensive tribute that was submitted by local people in Dunhuang to the khan of the Uighur Kingdom, along with jewels, silk, cotton, and hemp cloth. His study on economic documents from the cave library in Dunhuang also revealed that safflower and purple gromwell were the two most popular materials for making dyestuffs since most silk, woolen, and hemp textiles were dyed bright red or purple; here, safflower was used to produce a bright red color. Purple was often used for dyeing monastic robes to offer to high-ranking monks in the Dunhuang area. Safflower was also used for making rouge, a common cosmetic to wear on the cheeks for making them look bright red. Both textual and visual sources from Dunhuang illustrate the fashion of local women for wearing rouge. Nevertheless, it seems that processing technology limited the production of rouge, and it remained an expensive product from safflower, compared to its wide use for making dyestuff.

Trees and Wood Technology

In the Tang and Song Dynasties, one of the most remarkable accomplishments was the appearance of works on specific trees that illustrate meticulous knowledge about these trees. In particular, in the Song Dynasty, Chinese literati officials developed a predilection for planting trees and flowers in their private gardens and became interested in natural history. They were the first generation of Confucian scholars to focus their interest on nature rather than political and social order. Although the Chinese had already acknowledged the aesthetic and economic values of paulownia more than 3,000 years ago, a specific book on this plant did not appear until the eleventh century. Chen Zhu (1009–1061) was a native of Tongling County (modern Anhui Province). While being a hermit and without serving in the government as most Confucian scholars did, he devoted himself to cultivating bamboo and paulownia. He practiced the technology of cultivating paulownia and observed the growth of this plant. Eventually, in 1049, he wrote a book entitled *Record of Paulownia (Tongpu)* which offers a very detailed account of the history of this plant in China. He first traced the dispute on various names of this plant and identified them (including wutong (*Firmiaana simplex*), paulownia, and *tung* oil tree *Vernicia fordii*) by analyzing their tree forms, leaf forms, habitats, colors, flowers, and functions, then discussed its species. Some of these species can be identified more specifically by modern plant science. For example, from his description, the "white-flowered tong" (baihua tong 白花桐) is *Paulownia fortunei*. While the "purple-flowered tong" (zihua tong 紫花桐) is *P. tomentosa*. Chen Zhu focused on the extraction of oil from the seeds of *Vernicia fordii*. One of Chen's descriptions can also help identify *Erythrina variegata*, a tree that gave its name to Quanzhou as "Citong 刺桐" in the medieval period. Another species might be *C. japonicum*.[38] Chen Zhu also described the methods of planting, cultivating, maintaining, and processing. He provided a detailed account of the environmental conditions necessary for cultivating paulownia such as the required geomorphology, soil condition, sunshine, drying and soaking, temperature, and humidity. He even elaborated on a method for controlling harmful insects. According to his description, the planting of paulownia was very popular in the lower Yangtze River region (Figure 1.2). However, at that time, the most well-known

38 Stephen G. Haw, "Tung Oil and Tong Trees," *Zeitschrift der Deutschen Morgenländischen Gesellschaft* 167, no. 1 (2017): 224–25.

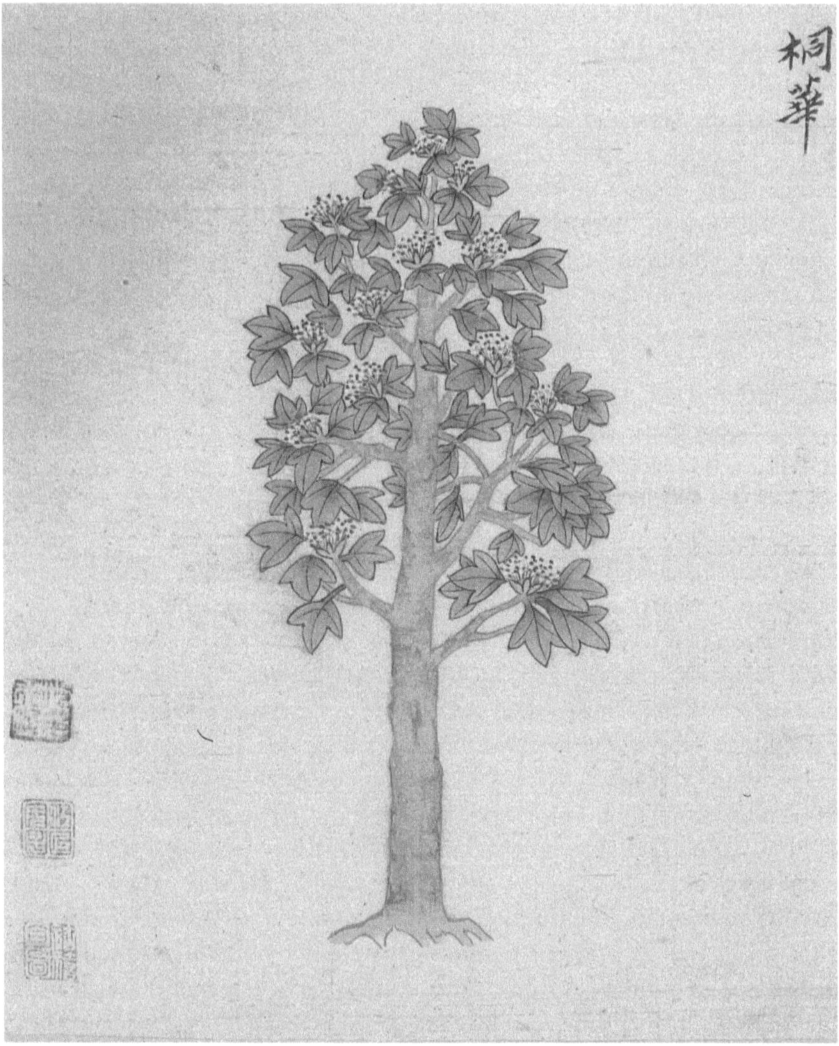

Figure 1.2: Painting of the paulownia from *Bencao tupu* (Illustrated Herbal) by Zhou Hu and Zhou Xi in 1644. Creative Commons (CC) license.

place for producing paulownia was the Sichuan area. He demonstrated his extensive knowledge of paulownia based on both his textual reading and personal experience.[39] Paulownia wood was widely used for various purposes, including making shoes and musical instruments. Chinese monks and nuns

39 Luo Guihuan, "Biology in Song China," in *Historical Perspectives on East Asian Science, Technology and Medicine*, ed. Chan Alan Kam Leung, Clancey Gregory K, and Loy Hui-chieh (Singapore: World Scientific, 2002), 173.

used wooden rice pots, bowls, and spoons.[40] According to Stephen G. Haw's study, two Chinese species *Vernicia fordii* and *V. montana* were important trees for extracting oil from their seeds. This *tung* oil was extensively used as paint and varnish in the Song and Yuan Dynasties. This oil became the chief paint oil for all external woodwork throughout the Chinese Empire in the early twentieth century. The oil was also used for making Chinese ink in the Song Dynasty.[41]

In the medieval period, it was noteworthy that *wooden architectural technology* flourished. Building science and technology in China experienced a sort of revolution and became much more mature and sophisticated than in previous dynasties; this can be marked by a meticulous manual entitled *State Building Standards (Yingzao fashi)*, a monumental work compiled in 1100 by Li Jie (1065–1110), a Superintendent of State Buildings, and released in 1103 by the Song government when Li Jie was promoted to Director of Palace Buildings. The appearance of this work should be attributed to the Wang Anshi Reform (1069–1074) launched by Wang Anshi 王安石 (1021–1086) who served as a chancellor for the socioeconomic reform movement and was an advocate for specialized professions. Wang urged the compilation of a guide to building construction. After Wang's death, the emperor commissioned Li Jie to formulate a new manual. This manual consists of thirty-four chapters divided into five parts, offering detailed technical guidelines for building palaces, mansions, ordinary houses, and pavilions. The first part interprets forty-eight building technicalities, mathematical formulae, building proportions, and information on construction. The second part lists standards and regulations for design, construction methods, working procedures, and manufacture of bricks, tiles, and glazed tiles. The third part focuses on work norms and costs as well as personnel management.[42] This manual provides highly reliable and specific information about the building technology in the Song Dynasty as well as previous dynasties. The major components required for constructing a building are blocks (dou 斗) and brackets (gong 拱), which means that timber was first processed for making these blocks and brackets. Flower-shaped bracketing seems to have been a typical technology from the ninth to the thirteenth centuries when wood-framed architecture extensively used brackets for structural and aesthetic purposes.[43] This particular wooden

40 Shiu-Ying Hu, "The Economic Botany of the Paulownias," *Economic Botany* 15, no. 1 (1961): 11–27.

41 Haw, "Tung Oil and Tong Trees," 215–36.

42 Qinghua Guo, "*Yingzao Fashi:* Twelfth-Century Chinese Building Manual," *Architectural History* 41 (1998): 1–13.

43 Jiren Feng, "Bracketing Likened to Flowers, Branches and Foliage: Architectural Metaphors and Conceptualization in Tenth to Twelfth-Century China as Reflected in the *Yingzao Fashi*," *T'oung Pao* 2nd Series, 93, no. 4/5 (2007): 369–32.

architectural technology which originated in medieval China demonstrated its power and also diffused abroad, to medieval Korea and Japan, especially in those traditional palaces and religious temples which are still in existence. This work also briefly mentions the following trees and woods for making various parts of medieval buildings: sandalwood (*Santalum album*), Siberian elm (*Ulmus pumila*), pagoda tree (*Styphnolobium japonicum*), pine (*Pinus*), cypress (*Platycladus orientalis*), zhennan or black wood (*Phoebe zhennan*), and Chinese fir (*Cunninghamia lanceolate*).

Architecture and Building

In particular, the earliest extant example of wooden architecture often refers to the East hall of Buddha's Light Temple and some nearby temple structures in Shanxi Province, which can be dated back to the ninth century. However, the original wood source is not always identifiable. Most likely, they were constructed of Prince Rupprecht's larch (*Larix gemlinii* var. *principis-rupprechtii*) and later other materials from different trees such as Chinese red pine (*Pinus tabuliformis*) and Chinese willow (*Salix babylonica [S. matsudana]*) were used for repair and replacement.[44] Interestingly, these trees can be found among the plants cultivated in Pingyuan Villa, the private garden of Li Deyu in the ninth century. It seems that in the Tang Dynasty, Chinese red pine and willow were commonly used for architectural purposes.

The wooden pagoda in Ying County, Shanxi Province, regarded as one of the oldest extant wooden buildings in China can be dated back to the eleventh century. The main parts of this building such as pillars, beams, arches, and blocks came from Prince Rupprecht's larch which is native to the Taihang Mountains in northern China, especially across the boundary between Shanxi and Hebei Provinces. Some other parts came from the Siberian elm and other pine trees.[45]

Shipbuilding

Shipbuilding technology experienced something of a renaissance in medieval China, becoming a more sophisticated and large-scale enterprise during

---

44 Tsinghua Daxue jianzhu sheji yanjiuyuan (Institute of Architectural Design at Tsinghua University), and Beijing Qinghua chengshi guihua sheji yanjiuyuan (Institute of Urban Planning and Design at Tsinghua, Beijing), *Foguangsi dongdadian jianzhu kancha yanjiu (An Examination and Study of the East Hall of Buddha's Light Temple).* (Beijing: Wenwu chubanshe, 2011), 49.

45 Hsueh-man Shen, *Gilded Splendor: Treasures of China's Liao Empire (907–1125)* (Milano: 5 Continents, 2006), 85–86.

this time. The construction of the great canal from Luoyang to Yangzhou and socioeconomic development in south China during the Sui and Tang Dynasties created a high demand for shipbuilding in Sichuan and other regions. Since the Tangut Empire blocked the Silk Route linking Central China and the Western Regions in the eleventh century, trade between China and other parts of Asia often took place by the sea, which provided the impetus for some major advances in shipbuilding technology in the Song Dynasty. After the Jurchen Dynasty occupied north China, the Southern Song Dynasty devoted itself to the building of a large navy to protect itself against enemies from the north along the Huai (across Anhui and Jiangsu Provinces) and Yangtze Rivers. Also, trade with Korea, Japan, and other regions needed to occur via sea; indeed, the Song Dynasty has been called the beginning of the maritime Silk Road era. In the Song Dynasty, technological innovations resulted in the construction of curved sideboards, broad lateral beams, and lofty superstructures with mortise and tenon joints and iron nails being employed to make boats stronger, safer, faster, and more powerful.

Both textual sources and archaeological discoveries provide some explicit information about shipbuilding technology in the Tang and Song periods. In the Tang Dynasty, most riverboats had flat bottoms and square prows. They were mostly made of zhennan or black wood (*Phoebe zhennan*), camphorwood, and Chinese fir (*Cunninghamia lanceolata*), as well as Chinese red pine (*Pinus tabuliformis*). Although these woods were the most commonly used in the shipbuilding industry, some small boats might have been constructed using other types of wood or other materials, depending on convenience and availability. Some fragments of canoes have been found in the Great Canal region near Liuzi (Huaibei, Anhui Province). Archaeologists identified one fragment, uncovered from Ningbo in Zhejiang Province in 1973, as a dragon boat from the Tang Dynasty. It was about 11.5 meters (38) long and 0.95 meters (3) broad, and cut from a single red pine timber. In the Song Dynasty, the most popular materials for building ships and boats were Chinese fir and Chinese red pine, which were widely available in south China. The Jiangxi region became the largest producer of riverboats and warships. Zhejiang and Fujian served as the principal places for building sea boats,[46] for there were abundant and good-quality wood materials available. For example, in 1974, the wreckage of a large sea boat from the Song Dynasty was uncovered in the Quanzhou Bay area, Fujian Province. The remaining structure of this boat was about 24.20 meters (79 feet) long and 9.15 meters (30 feet) broad. There are thirteen chambers inside. The keel is made of two red pine timbers.

46 Huang Chunyan, *Zaochuanye shiyu xia de Songdai shehui* (Society in Song China: A Perspective of Shipbuilding) (Shanghai: Shanghai renmin chubanshe, 2017), 16–18.

The cant frames were constructed from solid and durable camphorwood while other parts were made using Chinese fir as major materials. It has been found that in the process of construction, this boat was thoroughly painted with paulownia oil obtained from the seeds to solidify the structure.[47]

In medieval China, the government established numerous large shipbuilding yards for constructing riverboats, sea boats, and warships across south China, especially in the Jiangsu and Zhejiang areas. Some other places in Hubei, Jiangxi, Fujian, and Guangdong also had shipbuilding yards, producing small numbers of boats. As early as in the sixth century, the government built a large five-story warship which was nearly 55 meters (180 feet) long. It can be estimated that this would have had the capacity for 800 military personnel to defend it against enemies in the Yangtze River. In the Song Dynasty, a type of warship called the cart ship was equipped with two giant turning wheels on each side to power the ship. Four soldiers were required to move each wheel-board mechanical system. This warship could move much faster than its contemporaries. Some large warships might have been more than 100 meters (328 feet) long with 1500-ton capacities. In the early eleventh century, the government shipbuilding yards in south China produced more than nearly 3,000 riverboats and transport ships per annum. In 1257, in the Ningbo area alone, there were almost 8,000 boats. In one military campaign against the Jurchen invasion in the early thirteenth century, the Southern Song navy mobilized more than 700 warships. For trade purposes, boats made in China undertook voyages to Korea, Japan, India, Sri Lanka, and the Arabian Peninsula. Some sea boats were too large to enter the Persian Gulf, so merchants had to switch to smaller boats.[48]

Carts and Sedan Chairs

In terms of transportation in the medieval period, the Chinese tended to favor riding horses and using ox carts. A wooden model of an ox cart was discovered in an ancient tomb in Astana near Turfan, and was dated from the Tang Dynasty (618–907). It is only 20.8 inches (53 centimeters) high but clearly shows that an ox carried a black-painted cart for a single occupant.[49] However, in the Song Dynasty, sedan chairs made of wood and

---

47 Fujiansheng Quanzhou haiwai jiaotongshi bowuguan ed., *Quanzhouwan Songdai haichuan fajue yu yanjiu (The Discovery and Study on the Sea Boat in the Song Dynasty from the Quanzhou Bay)* (Beijing: Haiyang chubanshe, 2017), 21–22.

48 Feng Hanyong, "Tang Song shidai de zaochuanye (Shipbuilding in the Tang and Song Dynasties)," *Lishi jiaoxue* (*Teaching History*), no. 10 (1957): 10–14.

49 Liu Yonghua, *Zhongguo gudai cheyu maju (Carts, Carriages, and Horse Saddles in Traditional China)* (Shanghai: Shanghai cishu chubanshe, 2002), 146–48.

bamboo gradually became more popular for both officials and civilians. Images often depict carts and sedan chairs in the Tang and Song Dynasties, yet archaeologists rarely discover remains of medieval carts and sedan chairs, which makes it difficult to determine the original materials that were used to make them. Many sedan chairs seem to have been shouldered by two people and to have carried one occupant, though luxury sedan chairs for the upper classes might have required more people to lift. Pine and pear (*Pyrus* spp.) wood might have been used since they were popular materials for making furniture in the Song Dynasty. The famous painting *Along the River during the Qingming Festival* (*Qingming shanghe tu* 清明上河圖) by Zhang Zeduan 張擇端 (1085–1145) depicts numerous sedan chairs, and is commonly believed to reflect daily life in the capital city Kaifeng 開封 in the Song Dynasty.

Furniture in Medieval China

In the medieval period, grass, wood, and bamboo were often used to make furniture. Textual sources such as poems and prose from the Tang period mention a lot of furniture yet often do not specify the materials used for their construction. Medieval Chinese furniture can be categorized as follows: furniture for sleeping and crouching (beds and mats), furniture for sitting (chairs and stools), furniture for supporting objects and items (stands, tables, and desks), and furniture for storing objects and items (chests, boxes, cabinets, cupboards, and wardrobes). Archaeological findings of Tang furniture are very rare. However, some pieces preserved in Shōsōin 正倉院, a treasure house in Tōdai-ji 東大寺 in Nara, Japan, are believed to stem from the Tang Dynasty. They are made of narra (*Pterocarpus indicus*). According to the *Comprehensive Statutes* (*Tongdian* 通典) compiled in 766–801 by Du You 杜佑 (735–812 CE), many prefectures in North China (modern Shaanxi, Gansu) offered sabaigrass (*Eulaliopsis binata*) mats as local tributes to the Tang court, which seems to indicate its popularity in the northern regions. In other parts of China, sea water grass (*Cyperus malaccensis*) was used to make mats.[50]

In the Song Dynasty, various texts and archaeological discoveries provide information about furniture. Wood and bamboo continued to serve as two of the major materials for making furniture, though cane and grasses as well as other materials were also used. It appears that much furniture was made of wood from local trees,[51] and Chinese users recognized three

50 Liu Xianbo and Xiong Jun, *Tangdai jiaju yanjiu* (*A Study on Tang Furniture*) (Beijing: Renmin chubanshe, 2017), 33–34.

51 Shao Xiaofeng, *Zhongguo Songdai jiaju: Yanjiu yu tuxiang jicheng (Furnitures in Song China: Studies and Collected Images)* (Nanjing: Dongnan daxue chubanshe, 2010), 161.

wood categories: softwood (ruanmu 軟木), wood of inferior quality (chaimu 柴木), and hardwood (yingmu 硬木). The softwood in this, rather than the modern sense included Chinese aspen (*Populus adenopoda*), Chinese fir, paulownia (*Paulownia tomentosa*), and so on. In the Song Dynasty, Chinese aspen was commonly used for making screens, chests, boxes, and stands in North China. In the tomb of Yan Deyuan near Datong, Shanxi Province, illustrated screens and stands made of Chinese aspen were discovered. Chinese fir was a common tree in South China. A desk and a chair made of the wood of fir were found in the tomb of Sun Siniangzi in Jiangyin, Jiangsu Province, and dated from the Northern Song Dynasty. Paulownia was a popular choice for making musical instruments, such as Chinese zithers and drums.[52]

In the medieval sense, wood of inferior quality included Siberian elm (*Ulmus pumila*, jujube (*Ziziphus mauritiana*), Manchurian catalpa (*Catalpa bungei*), Chinese catalpa (*C. ovata*), apricot tree (*Prunus armeniaca*), and zhennan or black wood. Siberian elm and Manchurian catalpa commonly appeared as major materials for making furniture in the Yellow River region. The wood from these trees is durable in wet conditions, so it was also used for making agricultural tools. Manchurian catalpa was also a good choice for making Chinese go (weiqi 圍棋) game boards and scales. Chinese catalpa could be found in both the Yellow River and the Yangtze River regions, and was used for making tables, chests, and storage items, as well as the outer layer of coffins. Apricot tree wood mostly appeared in the Yellow River region. The chair, bed, screen, towel stand, table, and zither stand found in Yan Deyuan's tomb were all made of apricot wood. Southwest China was the major place for producing black wood, which was often used to make chests and coffins.

The hardwoods included sandalwood (*Santalum* spp.) and amboyna wood (*Pterocarpus indicus*), which were imported from southeast Asia and often used for making luxury furniture. One poem by the Northern Song Emperor Huizong seems to indicate that he possessed a sandalwood game board. A sandalwood container for storing Buddhist scriptures from the Northern Song was found in Rui'an, Zhejiang Province in the 1970s. Besides wood, bamboo was the second-most popular material for making furniture. There are numerous different species of bamboo in south China. Due to its easy availability and the less sophisticated processing technology, it was widely used for making everyday furniture, especially ordinary chairs, armchairs, beds, tables, mats, and sedan chairs. Various species of canes and grasses across the country were also used to make mats and stools.[53]

---

52 Shao, *Zhongguo songdai jiaju*, 163.
53 Shao, *Zhongguo songdai jiaju*, 161–68.

Woodblock Printing

In the Tang Dynasty, the Chinese invented the technology of woodblock printing. They used the hard, close-grained wood from pear (*Pyrus* spp.) and jujube (*Ziziphus* spp.) trees for making blocks.[54] The carvers first engraved Chinese characters on the blocks and then printed text by pressing paper on the blocks. It is well known that the earliest dated printed book can be traced back to the *Diamond Sutra*, dated 868, and discovered in the Dunhuang cave library. However, in reading historical sources, scholars have acknowledged that during the ninth century, some major urban centers of northern and central regions of the Tang Empire developed woodblock printed materials. For instance, in Sichuan, Chang'an, Luoyang, and Jiangxi areas, private printing activities might have produced medical books, poetry collections, reading primers, character lists, and divination books.[55] However, as early as the eighth century, some Buddhist seals, even if they were often not counted as books, might have been printed.

Exotic Plants from Abroad in Tang China

In the medieval period, the Tang Empire conquered many places in Central Asia and claimed numerous territories. As an open empire, it received various tributes from neighboring states and other regimes from afar. Persians, Indians, Sogdians, Turks, and Uighurs, as well as other peoples, often introduced exotic fruit to the Tang court. The Silk Road linking East Asia with West, South, and Central Asia became a pathway for constructing a Eurasian network of exchanges of religions, medicine, science and technology, animals and plants, as well as other materials. Many foreign plants were introduced into the Chinese Empire. In his monumental work entitled *The Golden Peaches of Samarkand: A Study of T'ang Exotics 618–907*, Edward H. Schafer discussed some of the most interesting exotic plants which were presented as royal gifts from abroad by drawing upon particular sources from Tang literature.[56] He also noted that some of these exotic plants later became common domestic plants in China. He first described the administration and technology of preserving fresh fruit such as litchis from Lingnan (modern Guangdong Province) and mare-nipple

---

54 Denis Twitchett, *Printing and Publishing in Medieval China* (New York: Frederic C. Bell, 1983), 70; Timothy H. Barrett, *The Women Who Discovered Printing* (New Haven: Yale University Press, 2008).

55 Joseph P. McDermott, *A Social History of the Chinese Book: Books and Literati Culture in Late Imperial China* (Hong Kong: University of Hong Kong Press, 2006), 11–12.

56 Schafer, *The Golden Peaches of Samarkand*, 117–32.

grapes from Central Asia (modern Xinjiang Province). The Chinese Empire established an office called the "Office of His Highness' Forest" in charge of receiving, transporting, and preserving perishable fruit, flowers, and seedlings from afar. This office ordered local administrators to cut ice blocks from the mountains for storing these perishable tributes. The imperial court of Tang China also built Tabooed Park near the imperial palace in Chang'an to serve as a nursery for planting exotic fruit and vegetables. A medical branch under the central government was responsible for planting medicinal plants. Some exotic plants from abroad discussed in Schafer's work include the date palm (*Phoenix dactylifera*) from Persia, peepul tree (*Ficus religiosa*) and sal tree (*Shorea robusta*) from India, saffron crocus (*Crocus sativus*) from India and Bukhara, snake flower (*Nagapushpam* in Sanskrit) from India, Narcissus from Persia, many species of lotus flowers from India, and water lilies (*Nymphaea caerulea*) from India. Although not noted by Schafer, many of these plants may also have been introduced by religious adherents of Zoroastrianism, Buddhism, Manichaeism, and Nestorianism since some of these plants served in liturgical rituals and daily practices. In Schafer's account, some exotic wood, mostly from India and Southeast Asia, also came to the Tang Empire. For example, sandalwood (*Santalum* spp.) was imported from India and Southeast Asia. Rosewood (*Dalbergia hainanensis*) was from the island of Hainan, and probably other rosewood came from Southeast Asia. A kind of blackwood from India, identified as ebony (*Diospyros* sp.) by Schafer, was also valued in Tang literature.[57]

Conclusions

The discussions above demonstrated that various plants have played sophisticated roles in the economic, social, cultural, and religious life of medieval China. The grains, vegetables, and fruits provided necessary material foundation for sustaining human lives. The agricultural technology enhanced in the medieval period helped support greatly increased population. The expanded cultivation of tea trees and textile plants made the economic and material exchanges within and beyond Asian continent possible and flourishing along the Silk Road. There were some remarkable landmarks in the historical development of plant science and technology in medieval China, which consisted of numerous new texts on the botanical knowledge and the invention of woodblock printing. Many of these innovations have reshaped the history of both plants and humans in Asian history.

57 Schafer, *The Golden Peaches of Samarkand*, 133–38.

Although this chapter has already covered a great number of plants, some plants are even more interesting. For example, honeysuckle (*Lonicerae japonicae*) was believed to be one of the most important medicinal plants in medieval Chinese herbology. It was commonly called "gold and silver flower (*Jinyinhua* 金銀花)" in medieval China due to its blossoming golden and white flowers. Along with chrysanthemum, its stems and flowers were used to reduce inflammation, fever, and heat cases of acute respiratory infection. More importantly, its pattern appeared as one of the most common decorative motifs in medieval Buddhist visual art, same as the palmette in classical art in Greece and Minor Asia which might have an impact on Indian and Buddhist art. In Japanese decorative art, it became well known as "the grass of Tang China (*kara kusa*)." In short, it was a typical international plant in medieval East Asia.

Chapter 2

ORDERING PLANTS IN THE BUDDHIST WORLD: A MEDIEVAL BOTANICAL TAXONOMY

Introduction

As the previous chapter briefly noted, plants were significant to the economic, cultural, and religious life in medieval China. Grains, vegetables, fruits, and raw materials for making textiles and papers helped sustain a society. Undoubtedly, plants played indispensable roles in the daily life of medieval Chinese Buddhist community. They provided material foundation for the daily operation of the Buddhist community, not only serving as food and nutrition nourishment, but also being presented as offerings to the Buddhist saints. In addition, plants often appeared as landscape decorations around the Buddhist sites, temples, and architectures. In other words, the Buddhist monastic and lay members developed both practical and symbolic meanings for having plants in their life. The Buddhist knowledge and concept of plants are also primarily based on the practical and spiritual needs of the Buddhist community, which emphasize moral and spiritual cultivations of the Buddhists, especially the monastic members.

In the long history of Buddhism from South Asia to China, similar to animals, some plants were more visible and remarkable for their close connections with the Buddhist tradition, such as the lotus flowers, for their affinity with the Buddha. While Asia is a vast land, different regions might technically cultivate and culturally value different plants. Even for traditional China, as a vast empire, many different regions cultivated their own indigenous plants. With the spread of Buddhism in China, the Buddhist community gradually grew and expanded, and more and more gardens and fields were donated by lay supporters to the monastic community.[1] The Buddhist community also received grains, vegetables, and fruits directly from the lay community. While dealing with both agricultural activities and plant offerings, some Chinese Buddhist masters, especially those who were involved in monastic administrations, developed their own knowledge and experience with plants in both Chinese and Buddhist traditions. This chapter will focus on how a medieval Chinese master Daoxuan (596–667) addressed the issues on the classification of plants in the monastic context. For me, this classification of the plants can be viewed as a Buddhist effort of ordering plants in medieval Buddhist world. This effort not only reflected a Buddhist master's knowledge about the botanical world, but also indicated how he reconciled the difference between Buddhist botanical knowledge and Chinese experience.

While serving as the superintendent of the Ximing Monastery, a leading royal temple in Chang'an, the capital city of the Tang Empire, Daoxuan wrote a series of concise brochures for offering his insights on ritual protocols and daily management of monastic life. Among these brochures, the *Ritual of Measuring and Handling Light and Heavy Property* (*Liangchu qingzhong yi* 量處輕重儀) was a text aiming at helping the monastic community deal with the property left by the deceased members. As a Buddhist master mainly being trained with the *Dharmagupta-Vinaya* (*Sifenlü* 四分律), he developed his rules and regulations by drawing upon resources from various traditions in both Buddhist and Chinese contexts. In this text, he offered his suggestions on how to handle animals and plants as monastic property in the Buddhist community.[2] This chapter will attempt to contextualize the list of plants as Daoxuan mentioned in his text.[3]

Botanical taxonomy has a long history in China.[4] In *Xunzi* 荀子, plants were classified into grasses and trees (*caomu* 草木). In the era of Warring

---

1 For a brief note on the plants in Chinese Buddhism, see Daniel Capper, *Roaming Free Like a Deer: Buddhism and the Natural World* (Ithaca: Cornell University Press, 2022), chapter 5: "Eating the Enlightened Plants of China"; for the study on container gardens in East Asian religions, see Rolf A. Stein, *The World in Miniature: Container Gardens and Dwellings in Far Eastern Religious Thought*, trans. Phyllis Brooks with a New Foreword by Edward H. Schafer (Stanford: Stanford University Press, 1990).

2 T. 45, no. 1895, 841a and 845b. For a brief discussion, see Huaiyu Chen, *The Revival of Buddhist Monasticism in Medieval China* (New York: Peter Lang Publishing, 2007), 148–51.

3 For the study on the list of animals, see Huaiyu Chen, *In the Land of Tigers and Snakes: Living with Animals in Medieval Chinese Religions* (New York: Columbia University Press, 2023), chapter 1.

4 Nguyen Tran Huan. "Esquisse d'une histoire de la biologie chinoise des origines jusqu'au IVe siècle," *Revue d'histoire des sciences* 10 (1957): 31–37; Joseph Needham, "The Development of Botanical Taxonomy in Chinese Culture," *Actes du douzième*

States, more sophisticated systems of classification appeared. For instance, in some early texts, plants were classified into one hundred types of flower (*baihui* 百卉), one hundred types of grain (*baigu* 百穀), one hundred types of vegetable (*baishu* 百蔬),[5] one hundred types of grasses (baicao 百草),[6] one hundred types of fruit (*baiguo* 百果), and one hundred types of medicine (*baiyao* 百藥).[7] The basis of these primitive systems of classifications is the forms and uses of the plants.[8] The first systematical classification of plants in ancient China appeared in *Erya*, a traditional lexicon. Based on classification of plants in *Erya*, some herbological and agricultural works developed the tradition of Chinese plants classification centered on medicinal plants.[9]

It should be note that some early medical texts also offered the classification of plants, thought they focused on medical plants. For instance, *Herbal Classic of the Divine Farmer* (*Shenong bencao jing* 神農本草經) that was allegedly compiled in the Western Han dynasty divided medical materials into three ranks, which herbs occupied a central position. The basic categories in this work included

---

congrés international d'histoire des sciences (1968): 127–33. For medicinal plants, see Hu Shiu-ying, *An Enumeration of Chinese Materia Medica* (Hong Kong: Chinese University Press, 1980). For an earlier survey of Chinese botany, see E. Bretschneider, "Botanicon sinicum. Notes on Chinese Botany from Native and Western Sources," *Journal of the North China Branch, Royal Asiatic Society* 16, 25, 29 (1881–1895), reprinted in Nendeln, Lichtenstein, 1967.

5 According to the *Discourses of the States* (Guoyu 國語), a legendary figure Lieshan 烈山/Lishan 厲山 once was a ruler and his son Zhu 柱 taught people how to cultivate hundred types of grains and hundred types of vegetables. Later the people in the Xia and Zhou dynasties honored Zhu as Houji 后稷 (the Lord Millet). See *Guoyu*, "luyu 魯語 1" (Shanghai: Shanghai guji chubanshe, 1978), 169–70.

6 See Guan Zhong 管仲, *Guanzi* 管子, ch. 57: On Appraising the Terrain (Duodi 度地). "Baicao" is a common word in early Chinese translations of Buddhist texts, along with five grains; see T. 1, *Chang ahan jing* 長阿含經, juan 2, 16a23; T. 1, *Chang ahanjing* 長阿含經, juan 18, 118a7; T. 20, *Fo kaijie fanzhi aba jing* 佛開解梵志阿颰經 (Pali: *Ambattha Sutta*), juan 1, 263a16; T. 197, *Foshuo xingqi xing jing* 佛說興起行經, juan 1, 166c12.

7 A common word appears in the *Record of Rites* (Liji 禮記), *Master Lü's Spring and Autumn Annals* (Lüshi Chunqiu 呂氏春秋), *Superfluous (chapters of the) Book of Zhou* (Yi Zhou shu 逸周書), *Huainanzi*, the *Book of Han Dynasty* (Han Shu 漢書), *Discussive Weighing* (Lunheng 論衡), and *Classic of Mountains and Seas* (Shanhaijing 山海經).

8 Zhongguo zhiwu xuehui ed., *Zhongguo zhiwuxue shi* 中國植物學史 (Beijing: Kexue chubanshe, 1994), 11–12. For study of crop plants, see Francesca Bray, "Essence and Utility. The Classification of Crop Plants in China," *Chinese Science* 9 (1989): 1–13. For earlier scholarship of evaluating ancient Chinese botanical works, see E. Bretschneider, "The Study and Value of Chinese Botanical Works," *Chinese Recorder* 3 (1870): 157–63.

9 Zhongguo zhiwu xuehui 中國植物學會 ed., *Zhongguo zhiwuxue shi* (Beijing: Kexue chubanshe, 1994), 12–14.

grasses, trees, fruits, grains, and vegetables.[10] This text was also well known to Buddhist scholars in ancient China. The first acclaimed translator An Shigao 安世高 already mentioned a text titled the *Herbal Classic* (*Bencao jing* 本草經) in his writings, which might refer to the *Herbal Classic of the Divine Farmer*, because this was the only available herbal classic at that time. Later Daoist scholar Tao Hongjing 陶弘景 (456–536) compiled a detailed commentary on this text, which was titled *Variorum of the Herbal Classic* (*Bencaojing jizhu* 本草經集注). In the early Tang Dynasty, *Newly Compiled Herbology* (*Xinxiu bencao* 新修本草), an official work sponsored by the central government in the Xianqing period of the Emperor Gaozong classified all plants into five categories based on their forms and uses: grasses (*cao*), trees (*mu*), fruits (*guo*), rice and grains (*mi gu*), and vegetables (*cai*).[11] Buddhists in the early Tang Dynasty, such as Daoxuan 道宣, seemed to have been aware of this tradition due to its release in AD 659. Huilin 慧琳 (737–820) has extensively cited *Newly Compiled Herbology* in his *The Pronunciation and the Meaning of All Scriptures*.

Although most monks did not engage in farming activities in medieval Chinese Buddhist monasteries, many plants were donated and cultivated. Buddhist masters faced the problem of dealing with these plants as monastic property. In order to determine the ownership of these plants, the masters first had to classify the plants and account for them. Unlike his economic classifications of laborers and animals, Daoxuan classified the plants in the Buddhist monastic communities mainly based on empirical principles. Daoxuan's system covers five kinds of vegetables, five fruit trees, and five grains.[12] It is worth noting that nowadays all these plants are considered vegetables.[13] Since they were

10 Wang Jiakui 王家葵 and Zhang Ruixian 張瑞賢, Shennong bencaojing yanjiu 神農本草經研究 (A Study of *Herbal Classic of the Divine Farmer*) (Beijing: Beijing kexue jishu chubanshe, 2001). There are also a number of articles in Chinese and Japanese languages dealing with both textual sources and topics in this text, which I cannot exhaust listing them here.

11 Zhongguo zhiwu xuehui ed., *Zhongguo zhiwuxue shi*, 38–43, esp. 41.

12 For plants in ancient India, see Trilok Chandra Majupuria, *Religious and Useful Plants of Nepal and India: Medicinal Plants and Flowers as Mentioned in Religious Myths and Legends of Hinduism and Buddhism* (Lashkar, 1988; revised by D. P. Joshi, 1989); Lambert Schmithausen, *The Problem of the Sentience of Plants in Earliest Buddhism* (Studia Philologica Buddhica, Monograph Series 6, Tokyo: The International Institute for Buddhist Studies, 1990), and his *Plants as Sentient Beings in Earliest Buddhism* (Faculty of Asian Studies, Australian National University, Canberra, 1991).

13 For economic plants in China, see Hu Xiansu, *Jingji zhiwuxue* 經濟植物學 (Beijing: Zhonghua shuju, 1953). Li Hui-lin, "The Vegetables of Ancient China," *Economic Botany*, 23 (1969): 253–60. For a study on the origin of cultivated plants, see N. I. Vavilov, *The Phyto-Geography Basis for Plant-Breeding, Origin and Geography of Cultivated Plants* (Cambridge: Cambridge University Press, 1992).

economically valuable, their significance in the monastic community was no less than in a non-monastic community. In other words, they served the daily needs of the monks. According to Daoxuan, the ownership of all plants belonged to the monastic community.[14] Individual monks should not own any plants.

As I have noted previously, Daoxuan's classification was based on his learning of *Four-part Vinaya*, and his classification was justified by the milieu in which he was situated. Indeed, some geographical and historical elements might have had an impact on Daoxuan's assessment of the monastic plants as they did on his assessment of the monastic animals. More specifically, Daoxuan's list illustrates that most plants, including vegetables and grains as well as fruit trees, were plants cultivated mostly in northern China where they were staple. Hence, Daoxuan's list reflected specifically the social-historical situation of the monastic community in northern China.

Why can we conclude that most plants in Daoxuan's list were cultivated in northern China? In taking a closer look at the plants Daoxuan discussed, we find that the terms he used were those common in northern China. In Daoxuan's system of classification, which adheres to the tradition of the Buddhist monastic codes, the five kinds of vegetables were ordered by five-birth types (*wusheng zhong* 五生種),[15] namely root (*genzhong* 根種), stem (*jingzhong* 莖種), joint (*jiezhong* 節種), miscellania (*zazhong* 雜種), and seed (*zizhong* 子種). This classification is slightly different from the classification system in the *Mahāsāṃghika-Vinaya*. In the *Mahāsāmghika-Vinaya*, there is a type called the type of mind (xinzhong 心種), which replaced "miscellania" in Daoxuan's list. The type of mind includes basil 蘿勒 (*Ocimum basilicum*, L.) and indigo 蓼藍 (*Persicaria tinctoria*, L.).[16] These terms appearing in the Chinese translations of early Buddhist texts might be different from the ones in Pali Buddhist texts. As Ananda Singh noted, there are two ways of classifying plants in India. In Indian religious literature plants are divided into seven categories, including medical herbs (*asadhi*), forest vegetation (vanaspati), trees with fruits and flowers (*urksa*), shrubs (*gumba*), grasses (*tina*), and plants with tendrils (*patāna*), and vines (*valli*). But in Buddhist literature, plants are identified on the basis of their roots (*mula*), stems (*khandha*), joints (*phulla*), head or uppermost pores (*agga*), or seeds (*bija*).[17] Singh is particularly concerned

14 Xie Chongguang 謝重光, "Jin Tang siyuan de yuanpu zhongzhiye 晉唐寺院的園圃種植業," *Zhongguo shehui jingjishi yanjiu* 中國社會經濟史研究 3 (1990): 1–7.

15 Li Hui-lin, "The Vegetables of Ancient China," *Economic Botany* 23, no. 3 (1969): 253–60.

16 *Mohe sengzhi lü* 摩訶僧祇律, juan 14, T. 22, no. 1425, 339a26–27.

17 Ananda Singh, *Planet, Plants and Animals: Ecological Paradigms in Buddhism* (Delhi: Primus Books, 2019), 84. Singh listed basil as one of the most sacred herbs in India. As an aromatic perennial plant, it has been known as holy basil appearing in many rituals (115).

with the moral attitudes in early Indian Buddhism toward plants. He pointed out that the monks, nuns, and lay people are encouraged to show friendliness and compassion toward the plants.

In Daoxuan's list, the type of root includes radish (*luobo*) (Figure 2.1),[18] ginger (*jiang* 薑),[19] Colocasia antiquorum (*yu* 芋),[20] and so on. Ginger was an important vegetable for ascetic practice in Buddhist culture. In *Pāyāsisuttanta* (*Pisi jing* 蜱肆經) collected in *Madhyamāgama* (*Zhong Ahan jing* 中阿含經), it said that, when the King Pāyāsī (Pisi wang 蜱肆王) began to cultivate his merits, he only ate a piece of ginger for practicing his asceticism.[21] Also in *Madhyamāgama*, it said that a layman Sudatta always ate vegetables, including a piece of ginger. In *Mahāparinirvāṇa-Sūtra*, ginger was said to heal the ill people who caught cold. In the Ten-stanza Vinaya, ginger was listed as one of five root medicines, which means the roots of five materials were used in Buddhist medicine. Other four materials included Śarīra (relics), Radix Aconiti Lateralis Preparata (fuzi 附子),[22] Prati-visa (more commonly known as Ativisa, modern Latin name: Aconitum heterophyllum, Ch. Boti pisha 波提毗沙), and Acorus calamus (changpu 昌蒲).[23] Xuanzang in his *Record of the Western Regions of the Great Tang* (*Da Tang xiyu ji* 大唐西域記) also noted ginger was one of the crucial vegetables in India. Some biographies have indicated that monks

---

18 Latin: *Raphanus sativus*. For a short history of its cultivation, see Li Fan, *Zhongguo zaipei zhiwu fazhanshi*, 107–9.

19 Latin: *Zingiber officinale*. Li Fan suggests that this plant was originally cultivated in western plateau of China, especially in present Yunnan and Guizhou areas. See his *Zhongguo zaipei zhiwu fazhanshi*, 132–33. It was well cultivated in ancient India and China; see Berthold Laufer, *Sino-Iranica: Chinese Contributions to the History of Civilization in Ancient Iran, with Special Reference to the History of Cultivated Plants and Products* (Chicago: Field Museum of Natural History, 1919); see Chinese version, Lin Yunyin trans. *Zhongguo Yilang bian: Zhongguo dui gudai Yilang wenmingshi de gongxian zhuozhong yu zaipei zhiwu ji chanpin zhi lishi* 中國伊朗編：中國對古代伊朗文明史的貢獻，著重于栽培植物及產品之歷史 (Beijing: Shangwu yinshuguan, 1964, reprinted in 2001), 376–77.

20 Latin: *Colocasia esculentum*. For its cultivation in China, see Su Song, *Tujing bencao* (Fuzhou: Fujian kexue jishu chubanshe, 1988), 488–89.

21 *Zhong Ahan jing*, juan 16, see *T.* no. 26, vol. 1, 532a.

22 Yu Xin 余欣, "Fuzi kao: Yaowu de dongxi jiaotong shi 附子考:藥物的東西交通史 (A Study on Fuzi: The Medical Exchanges between the East and West)," in his book *Zhonggu yixiang: Xieben shidai de xueshu, Xinyang yu Shehui* 中古異相：寫本時代的學術、信仰與社會 (Mapping the Extraordinary: Knowledge, Faith and Manuscript Culture in Medieval China) (Shanghai: Shanghai guji chubanshe, 2011), 189–216; Dandan Zhao, Jie Wang, Yanjing, Cui, and Xinfang Wu, "Pharmacological Effects of Chinese Herb Aconite (Fuzi) on Cardiovascular System," *Journal of Traditional Chinese Medicine* 32, no. 3 (2012): 308–13.

23 Stephen R. Bokenkamp, "The Herb Calamus and the Transcendent Han Zhong in Taoist Literature," *Studies in Chinese Religions* 1, no. 4 (2015): 293–305.

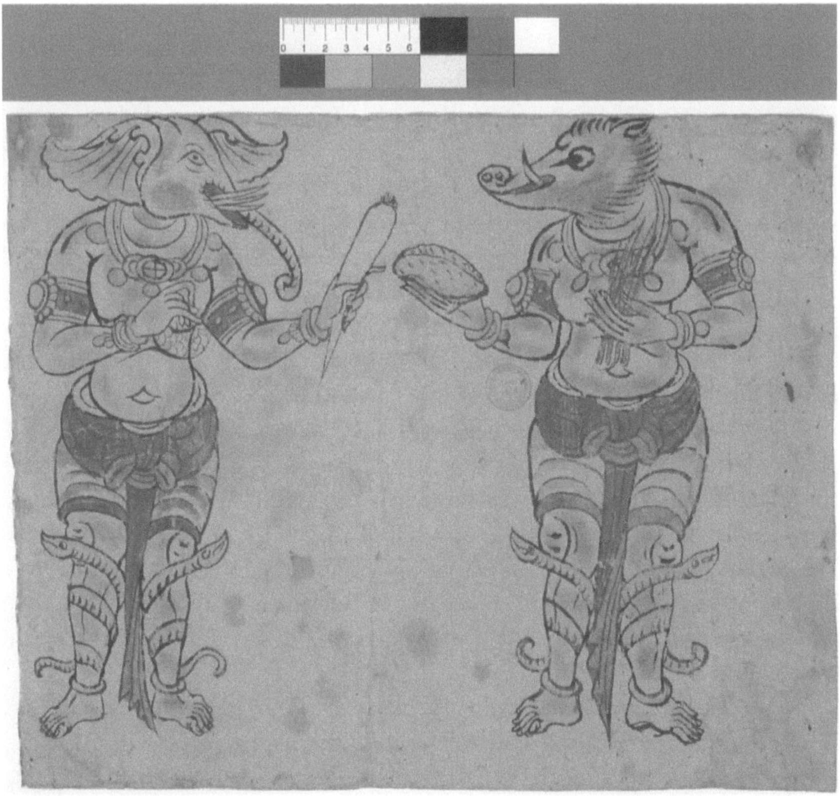

Figure 2.1: Dunhuang manuscript Pelliot chinois 4518v8 shows the images of South Asian gods Ganeśa and Varaha and the former holds a radish in his left hand. © The Bibliothèque nationale de France.

in ancient China ate ginger in their practice. For instance, Shan Daokai 單道開, a native of Dunhuang, was said to restrain himself from grains. He often ate the fruit of Platycladus orientalis (bai 柏), pine rosin (Songzhi 松脂), even small stones, and also ginger, for seven years.[24]

The type of stem includes pomegranate (*liu* 榴), willow (*liu* 柳), and some similar vegetables. In medieval Chinese historical sources, pomegranate sometimes was also called "an shiliu 安石榴," which refers to the pomegranate from Bukhara, because the Chinese called Bukhura as the Kingdom of An, one of Sogdian oasis city states in Central Asia. In Huilin's *The Pronunciation and the Meaning of All Scriptures*, he cited sources from Zhang Hua's (AD 232–300) *Records about Vast Topics* (Bowuzhi 博物志), this plant was brought

24 Huijiao, *Gaoseng zhuan*, juan 9, see *T.* no. 2059, vol. 51, 878a.

back along with grape and peach from Persia by Zhang Qian to the Central Plain in the Western Han dynasty. The pomegranate was not well planted in China and rarely appeared in Chinese Buddhist historical sources. It occasionally appeared in traditional official histories. For example, according to the biography of Zhang Chang 張暢 in the *History of Southern Dynasties* (*Nanshi* 南史), this fruit was planted in the Ye 鄴 area.[25] When Chinese pilgrim Faxian 法顯 went to India, he noticed that three plants across both sides of the Pamir plateau were same, which means that they all appeared in both the Central Plain and the Western Regions. These three plants included bamboo, pomegranate, and sugarcane.[26] When Xuanzang passed by Kucha, he found that this area produced grape, pomegranate, peach, pear, and apricot, as well as other fruits.[27]

The type of joint (jiezhong) includes basil (*Ocimum basilicum*, Ch. *luole* 蘿勒), polygonum (*liao* 蓼), Chinese celery (*qin* 芹).[28] Basil also appeared in Daoxuan's commentary on the *Caturvargika-Vinaya*. According to Daoxuan's Continued Biographies of Eminent Monks, an early Tang monk, Xuanyi 玄嶷 moved Wu Zetian by keeping himself eating polygonum as an ascetic practice. We Zetian gave him green light in converting to Buddhism from Daoism. He then had his head shaved and resided in the Temple of the Buddha's Prediction (Fo shouji si 佛授記寺).

The type of *miscellania* includes sugarcane (*zhe* 蔗),[29] bamboo (*zhu*), *Anaphalis yedoensis* (*di*), and rush (*Phragmites communis*, Ch. *lu*). In 1061, Su Song finished compiling a herbal work titled *Illustrated Scripture of Herbs* (*Tujing bencao*) and listed sugarcane along with persimmon and apple as types of fruit.[30] It is also worth noting that in Su Song's classification, fruits include

25 Li Yanshou, *Nan Shi*, 32. 22 (Beijing: Zhonghua shuju, 2011), 831. According to the *Book of the Southern Qi Dynasty* (Nanqi shu 南齊書) (58.39), in the Kingdom of Funan 扶南 (Hvunân, Phù Nam), there were a lot of sugarcanes, circus, pomegranate, and areca palm, see Xiao Zixian, *Nan Qi shu* (Beijing: Zhonghua shuju, 1987), 1017.

26 *Gaoseng Faxian zhuan* 高僧法顯傳, T. 51, no. 2085, 857c24. Also See Liu Zhaomin 劉昭民 and Liu Tangrui 劉棠瑞, *Zhonghua shengwuxue shi* 中華生物學史 (*A History of Chinese Biology*) (Taipei: Taiwan shangwu yinshuguan 1991), 196.

27 Xuanzang, *Da Tang xiyu ji* 大唐西域記, T. 51, no. 2087, 870a17. He called Kucha as "Quzhi guo 屈支國."

28 According to Li Fan, there are two kinds of celeries: *Oenanthe javanica* and *Apium graveolens*. It seems that the celery Daoxuan lists here is the former, which was more popular as vegetable cultivated in ancient China. For more information about these two celeries in ancient China, see Li Fan, *Zhongguo zaipei zhiwu fazhanshi*, 138–39.

29 *Xinxiu bencao* 新修本草 (ch. 17) cites Tao Hongjing's *Bencaojing jizhu* 本草經集注 that this plant was mostly cultivated in southeastern China. Su Song also gives a brief description of this plant. See Su Song, *Tujing bencao*, 487–88.

30 Su Song, *Tujing bencao*, 487–88.

both *Colocasia antiquorum* (*yu*) and sugarcane (*zhe*), which appear as vegetables in Daoxuan's classification. Some biographies in Daoxuan's *Continued Biographies of Eminent Monks* have told that in medieval period, Chinese monastic members did eat these vegetables. For example, Shiyu 世瑜, a monk who went forth in AD 616 observed the Dhuta, an ascetic Buddhist practice. He entered the Mount Xiangying (Xiangying shan 響應山, modern Mianzhu 綿竹 in Sichuan province) and lived alone. Apes offered him mountain fruits for keeping alive. Some laymen also provided him the lotus flower, sugarcane, and taro (Colocasia antiquorum). So for three years while practicing in mountain, he only consumed rice for seventeen pecks.[31]

In Daoxuan's list, the fifth group of vegetables, or the type of seed included coriander (*sui* 荽, *Coriandrum sativum*. L) and Perilla ocimoides (*ren* 荏).[32] Coriander herb was introduced from the Western Regions by Zhang Qian in the Western Han dynasty, so it was commonly called "barbarian coriander herb" (husui 胡荽) in the Central Plain. In the section of Buddhism and Daoism (*Shilaozhi* 釋老志) of the *History of Wei Dynasty* (*Weishu* 魏書), it said that a Daoist practitioner Wang Daoyi 王道翼 lived as a hermit in the Mount Hanxin (Hanxin shan 韓信山, modern Jingjing 井陘 in Hebei province) for more than forty years, restraining himself from grains but relying on this Coriander Herb.[33] Interestingly, among Dunhuang manuscripts a text titled *On Essentials of Opening the Meaning of Tradition in Mahāyāna Brilliant Gates and Hundred Methods* (Dasheng baifa mingmen lun kai zongyi jue 大乘百法明門論開宗義決) preserved the information about how people in different regions called this plant. According to this text, in the Hexi Corridor, people called it "barbarian" coriander herb (husui 胡荽). In Shandong area, it was called coriander herb (Sui 荽). And, in the Lower Yangtze River and the Huai River area it was called the fragrant coriander herb (Xiangsui 香荽).[34] This text was attributed to Tankuang 曇曠, a monk in the Tang dynasty. Although it was found in Dunhuang, as a monk in North China, Daoxuan seemed not to be familiar with the name of this plant that was used in the South.

Coriander was not regarded as one of five forbidden vegetables in Chinese Buddhism until in the post-Tang period. For instance, in the biography of Huiri 慧日 (68–748),[35] it said that there were five forbidden spicy vegetables (Skt. parivyaya, Ch. wuxin 五辛) in India, such as garlic, green onion, leek, *Allium chinense*, Asafoetida (in Fayun's *Fanyi mingyi ji*),

31 Daoxuan, *Xu gaoseng zhuan* 續高僧傳, juan 20; *T.* no. 2060, vol. 50, 595a.
32 Daoxuan, *Liangchu qingzhong yi* 量處輕重儀, *T.* no. 1895, vol. 45, 841a.
33 Wei Shou, *Weishu*, juan 114, modern annotated edition (Beijing: Zhonghua shuju, 1974), 3054–55.
34 Tankuang, *Dasheng baifa mingmeng lun kaizongyi jue*, see *T.* no. 2812, vol. 85, 1086c.
35 Zanning, *Song Gaoseng zhuan*, juan 29, see *T.* no. 2061, vol. 50, 890b.

or Coriander herb.[36] These vegetables were believed to arouse or disturb the emotional responses that would corrupt morality and spiritual progress of the Buddhist practitioners. In early Buddhist monastic code, monks and nuns were not allowed to consume garlic as a vegetable, but they could eat it as medicine. Yet in China, there were only four of these five forbidden vegetables, and the Coriander Herb was not one of them.

As for fruits, Daoxuan's list also shows that most fruits he lists were cultivated at his time in northern China. Daoxuan has listed two kinds of classification for fruits. The first category is called five fruits[37] presently on the tress (*xianshu wuguo* 現樹五果). They are fresh fruits that included the fruit of shell (*keguo* 殼果), the fruit of skin (*fuguo* 膚果), the fruit of core (*hegu* 核果), the fruit of horn (*jiaoguo* 角果), and the fruit of cart (*yuguo* 輿果).[38] According to Daoxuan, the first group of fruits includes walnut (*hutao* 胡桃) and chestnut (*li* 栗).[39] *Hutao* also appears in Fayun's *Collection of Names and*

---

36 The Mahayana Buddhist precept text *The Brahmajāla Sūtra* (Fanwangjing 梵網經) clearly stated that all Buddhists should not eat five spicy vegetables; see T. 24, no. 484, 1005b14. In many Buddhist texts such as the Chinese versions of the *Mahāparinirvāṇa Sūtra* and *Mahāsaṃnipātasūtra* translated by Dharmakṣema, five spicy vegetables were listed with meat and alcohol drinks together and they were all prevented from consuming by the monastic members (juan 11); see T. 12, no. 374, 432c. In medieval China, it seems that this rule was widely accepted, by a lot of people, including those who held Confucian values. At least in the Tang dynasty the filial son might also abstain from five forbidden spicy vegetables, which seems to indicate the Confucian acceptance of Buddhist value and practice or a popular combination with both Confucian and Buddhist values. For instance, Wu Zaide 武載德, the grandson of Empress Wu Zetian's uncle Wu Shiyi 武士逸, observed the rule of abstaining himself from eating five vegetables after his older sister passed away; see Ouyang Xiu et al, *Xin Tang shu* 新唐书, 206, 131 (Beijing: Zhonghua shuju, 1975), 5838.

37 Skt. *Pañca-phalāni*. This term "wuguo" did not appear in Chinese official histories until Western Jin Dynasty (fourth century); see Chen Shou, *History of Wei* (*San guozhi: Wei shu* 三國志魏書), ch. 16, 511, 841. While in Wei Shou's *History of Northern Wei* (*Wei shu*), it was used to describe the situation in Western Regions (*xiyu*); see *Wei shu*, 2222, 2243, 2272–74, 2278. Similar uses can be found in *History of Northern Dynasties* (*Bei shi*), 915, 3131, 3212, 3224–25. It might be borrowed from Buddhist sources.

38 The name of five fruits appears in *Foshuo yulanpen jing* 佛說盂蘭盆經, trans. Zhu Fahu (third century) see *T.* no. 685, vol. 16, 779b. While modern monk Cizhou comments on *Foshuo yulanpen jing* and identifies these five fruits as *keguo*, *heguo*, *fuguo*, *huiguo*, and *jiaoguo*. For a discussion on this scripture, see Stephen F. Teiser, *Ghost Festival in Medieval China* (Princeton: Princeton University Press, 1988), 48–56.

39 This classification system is also different from that in the *Mahāsāṃghika-Vinaya*, juan 14; see T. 22, no. 1425, 339b11. In this section, there are the following five types of fruits: type of covering the core裹核種、type of covering with skin 膚裹種、type of covering with shell 觳裹種、type of covering with husk 檜裹種、and type of covering with horn 角裹種.

Meanings in Translations (*Fanyi mingyi ji* 法門名義集). In this Buddhist work, it was suggested that its Sanskrit form is *Juglans regia* (ch. 24). However, based on its transliteration *pārasī* which means "Iranian," Berthold Laufer argued that this fruit originally came from Iran.[40] He also noted that in the fourth century, not many Chinese people knew of this plant. He cited Su Song (1020–1101) in *Illustrated Scripture of Herbs* (*Tujing bencao*) and says that *hutao* seems to be cultivated mostly in northern China.[41] In Xuanzang's *Record of the Western Regions in the Great Tang*, it said that in Tocharian state, many fruits including "hutao," grape, pear, and apricot were planted.[42] The *Dharmagupta-Vinaya* said that the monastic members were allowed to consumer grapes, hutao, pear, and pineapple.[43] Daoxuan accepted this rule in his commentary on the precepts for monks based on the *Dharmagupta-Vinaya*.[44] According to Guanding 灌頂 (561–632), when Yang Guang was a crown prince, he offered "hutao" to Zhiyi 智顗 (538–597), the acclaimed Tiantai master.[45] It might be due to the fact that "hutao" was a rarely seen fruit in South China. Yang Guang attempted to show his respect to Zhiyi, so offered him such precious fruit from the Western Regions.

The second group of fruits includes pear (*li* 梨),[46] crab apple (*nai* 柰), apple (*linqin* 林檎, *Malus pumila*, L.),[47] and melon (*mugua* 木瓜).[48] The third group of fruits includes peach (*tao* 桃), apricot (*xing* 杏, *Prunus armeniaca*, L.),[49] jujube (*zao* 棗),[50] and persimmon (*shi* 柿).[51] The fourth group of fruits includes all

---

40 See his *Sino-Iranica: Chinese Contributions to the History of Civilization in Ancient Iran, with Special Reference to the History of Cultivated Plants and Products*, 254–56.

41 Ibid., p. 257. Su Song's book is an excellent to trace where the plants were cultivated. Also see Su Song, *Tujing bencao*, 494–95.

42 Xuanzang, *Da Tang xiyuji*, juan 12, see *T.* no. 2087, vol. 51, 940b.

43 *Sifen lü* 四分律, juan 21, see T. 22, no. 1428, 707c17–8. According the early Buddhist text *Dafangguang sanjie jing* 大方廣三戒經 translated by Dharmakṣema, many of trees in the Vulture Peak produced numerous fruits such as pear, peach, apricot, grapes, hutao, pomegranate, etc.; see T. 11, no. 311, 687a20–22.

44 Daoxuan, *Sifen lü biqiu hanzhu jieben* 四分律比丘含注戒本, T. 40, no. 1806, 459a15–16.

45 Guanding, *Guoqing bailu* 國清百錄, juan 3, see *T.* no. 1934, vol. 46, 814b.

46 Su Song, *Tujing bencao*, 496–97.

47 Both *nai* and *linqin* are archaic names of apple. According to Li Fan, *nai* is a little bit larger than *linqin*, and both these two fruit trees were mostly cultivated in northwest China; see his *Zhongguo zaipei zhiwu fazhanshi*, pp. 178–79. In modern Japanese, kanji "*linqin*" refers to apple.

48 Su Song, *Tujing bencao*, 486–87.

49 It was cultivated mostly in north China. See Li Fan, *Zhongguo zaipei zhiwu fazhanshi*, 172–74.

50 It was originally cultivated in northwest and north China. See Li Fan, *Zhongguo zaipei zhiwu fazhanshi*, 196–98; Su Song, *Tujing bencao*, 497–98.

51 Su Song says that it was cultivated in both south and north. See Su Song, *Tujing bencao*, 485–86.

kinds of beans cultivated in valleys and marshes (*shanze zhudou* 山澤諸豆). The final group of fruits includes the seeds of pine tree and cedar tree (*songbaizi*). The second category indicates the dried fruits, including dried jujube, pear, plum (*li, Prunus salicina*, L.),[52] apricot, persimmon, large tangerine (*gan* 柑), tangerine (*ju* 橘), orange (*cheng* 橙), chestnut (*li*),[53] and pomelo (*you* 柚). Many monks were said to rely on these fruits while pursued their Buddhist practice alone in the mountains. For instance, a monk Shi Sengcong 僧從 lived in the Mount Pubu in Shifeng 始豐 area, only took jujube and chestnut, without eating five grains. Even he was nearly one hundred years old he was still strong and not tired of reciting Buddhist scriptures (*Gaoseng zhuan*, juan 11). In early Tang period, Huijue 慧覺 planted jujube trees in the Qing prefecture (Qingzhou 青州 in Shandong area) for all five assemblies of Buddhists (*Xu gaosengzhuan*, juan 12). When Yang Jian 楊堅 was the prime minister in the Northern Zhou dynasty, he offered ten pecks of jujube from Qingzhou to Fazang 法藏 (*Xu gaoseng zhuan*, juan 19). Other eminent monks were also said to live on fruits, such as jujube, chestnuts, and some others. Given the nature of these biographies as hagiographies for making the images of eminent monks, these stories recorded in these biographies have still more or less shown some aspects of monastic life in forests and mountains. Daoxuan also noted that the auspicious place might produce tangerine, pomelo, and plums. For example, there was a square pond in a valley near Yongzhou (modern Hunan Province) and a stupa for housing the Buddhist relics in the south of the pond, while on the left side of the valley there were numerous trees of tangerine, pomelo, and plums.[54] In the Tang Dynasty, with the spread of the medieval Chinese ritual tradition, the pomelo was a common fruit used in the esoteric ritual, as it appeared to be held by one of the right hands of Mahāmāyūrī. Another right hand held a lotus flower. On the left side, two hands held bael and three pieces of peacock tail feathers.[55]

It should be discussed here that as a monk who grew up in North China, Daoxuan does not mention some typical fruit trees that have been long cultivated in southern China, such as lychee (*lizhi* 荔枝),[56] banana (*xiangjiao*

52 See Li Fan, *Zhongguo zaipei zhiwu fazhanshi*, 169–70.

53 Li Fan, *Zhongguo zaipei zhiwu fazhanshi*, 199–201.

54 Daoxuan, *Lüxiang gantong zhuan* 律相感通傳 (Narrative of Spiritually Miraculous Transmissions on Matters of Monastic Discipline), T. 45, no. 1898, 879a3–4.

55 Yijing trans., *Foshuo Da kongque zhou wang jing* 佛說大孔雀呪王經 (Scripture Spoken by the Buddha on the Spells of King Mahāmāyūrī), juan 3; see T. 19, no. 985, 476b24–25.

56 Su Song, *Tujing bencao*, 489–90. Su Song indicates that litchi from Sichuan area was particularly famous in the Tang Dynasty, based on Bai Juyi's work. Now Bai Juyi's work is not extant. Su Song also says that at his time, the litchi from Fujian was better than others.

香蕉),[57] olive (*ganlan* 橄欖),[58] and coconut (*yezi* 椰子).[59] Lotus root, lotus seed, and water caltrop that have appeared in the *Dharmagupta-Vinaya* and the *Ten-Stanza Vinaya* frequently were also missed in Daoxuan's list. While their names did not appear in Daoxuan's work, these fruits and their history in southern China were recorded in many medieval works. For example, the *Guide of Grasses and Trees in the South* (*Nanfang caomu zhuang* 南方草木狀)[60] has mentioned that lychee, coconut, banana, and olive were cultivated in south China.[61] Coconut appeared with "hutao" and pomegranate together in the *Mahāsāṃghika-Vinaya*. Xuanzang also noted in his *Record of the Western Regions in the Great Tang* that the people in Nārikela (modern Maldives) ate coconut as main food in daily life. As the name of a state, Nārikela was actually the Sanskrit name for coconut. Daoxuan was familiar with this source and borrowed it in his the *Śākya Gazetteers* (*Shijia fangzhi* 釋迦方志). Huilin said in his *The Pronunciation and the Meaning of All Scriptures* that coconut was a common fruit in Guangzhou area. Daoxuan did not mention it perhaps that in North China it hardly appeared as a crucial fruit to local monastic community.

Lycee was another tropical fruit missing from Daoxuan's list.[62] As early as in the Western Han dynasty, this tropical fruit has been known to the people living in the Central Plain, as the *Rhapsody on the Shanglin Park* (*Shanglin fu* 上林賦) by Sima Xiangru 司馬相如 (179–118 BCE) has shown. Regarding to the Buddhist literature, it also appeared one of the translations of Dharmarakṣa (Zhu Fahu 竺法護), which is the *Śārdūlakarnāvadāna*. In this text, it discussed the four forms of births for sentient beings and also the types of plants. The knowledge of plants was based on the roots,

57 In India, as Singh noted, "Banana stems, leaves, and fruits are used in many auspicious occasions and rituals"; see *Planet, Plants and Animals*, 106.

58 Su Song, *Tujing bencao*, 499–500.

59 For coco, see Li Fan, *Zhongguo zaipei zhiwu fazhanshi*, 223–24.

60 For an English translation, see Li Hui-lin 李慧林, trans. *Nan-fang ts'ao-mu chuang. A Fourth Century Flora of Southeast Asia* (Hong Kong: Chinese University Press, 1979). Ma Tai-loi examines its authenticity; see his "The Authenticity of the *Nan-fang ts'ao-mu chuang*," *T'oung Pao* 64 (1978): 218–52.

61 Li Fan, *Zhongguo zaipei zhiwu fazhanshi*, 218–19; 223–27. *Miscanthus sinensis* (*mangguo*) was also brought by Xuanzang to China in seventh century; see Li Fan, *Zhongguo zaipei zhiwu fazhanshi*, 228.

62 For the study of lycee in Chines history, see Luo Guihuan 羅桂環, "Woguo lizhi de qiyuan he zaipei fazhan shi 我國荔枝的起源和栽培發展史," *Gujin nongye* 古今農業, 3 (2001): 71–78; Zhou Zhaoji 周肇基, "Lidai lizhi zhuanzhu zhong de zhiwuxue shengtaixue shenglixue chengjiu 歷代荔枝專著中的植物學、生態學、生理學成就," *Ziran kexue shi yanjiu* 自然科學史研究 10, no. 1 (1991): 35–47; Chen Yuanpeng 陳元朋, "Lizhi de lishi 荔枝的歷史," *Xin shixue* 新史學 14, no. 2 (2003): 111–78.

stems, branches, leaves, flowers, and fruits. It listed several fruits such as pear, apricot, jujube, chestnut, melon, cherry, hutao, longan, lychee, grape, etc.[63] In official histories, it was also viewed as an important local product that would be offered to the central government, the Tang court. In the eighth century, lychee was the favorite fruit of the emperor Xuanzong 玄宗 (Li Longji) and his concubine Yang Guifei 楊貴妃. Many poems in the Tang dynasty have discussed this fruit against the historical background about Yang Guifei and her lover Li Longji, the Brilliant August (Minghuang 明皇) of the Tang dynasty, and the An Lushan rebellion that brought the Tang Empire to decline. The Famous poet Bai Juyi 白居易 (772–846) wrote a preface for a work titled the *Illustrations of Lycee* (Lizhitu 荔枝圖).[64] Another famous poet Du Mu 杜牧 (803–852) left a famous poem titled *On Passing by the* Hua-qing *Palace* (Guo Huaqing gong 過華清宮) about Yang Guifei's appetite of lychee and its relationship with the An Lushan rebellion.[65] Finally, it should also mention the lychee in East Asian Buddhist context that, when Japanese monks Eiei 榮叡 and Fushō 普照 came to China, they were offered lycee, longan, and sugarcane.[66]

Another famous tropical fruit Daoxuan did not mention in his list is mango, is called *āmra* in Sanskrit (Pali: *amba*) and called *anluo* 庵羅 (*am-la* or *am-ra*) in Chinese.[67] Mango is regarded as a divine fruit in India, which has healing power and has been used as ayurvedic medicine. The tree of mango is considered auspicious. Some early Buddhist texts even mentioned that the Buddha sometimes delivered his sermons in the mango groves.[68] Looking at Chinese Buddhist literature, it appeared very often in numerous texts, such as *Saṁyuktāgama*,[69] *Mahā-parinirvāṇa-sūtra*,

63 Zhu Fahu tr., *Shetoujian taizi ershibaxiu jing* 舍頭諫太子二十八宿經 (*Śardūlakarnāvadāna*), see *T.* no. 1301, vol. 21, 413b.

64 See Ouyangxiu et al., *The New History of the Tang Dynasty* (*Xin Tang shu* 新唐書), juan 155, modern annotated edition (Beijing: Zhonghua shuju, 1982), 4352.

65 For an English translation, see Stephen Owen, *An Anthology of Chinese Literature: Beginnings to 1911* (New York: W. W. Norton & Company, 1997), 452.

66 Genkai 元開, *Tō daiwajō tosei den* 唐大和尚東征傳, see *T.* no. 2089, vol. 51, 991a.

67 Berthold Laufer, *Sino-Iranica*, 552. Its Latin name is known as mangifera indica. Laufer also notes that mango was a native Indian fruit, called the king of Indian fruits. It is also considered as one of the most delicious fruits in ancient India, see Lambert Schmithausen, *The Problem of the Sentience of Plants in Earliest Buddhism* (Tokyo: The International Institute for Buddhist Studies, 1991), 37.

68 Singh, *Planet, Plants and Animals*, 101.

69 For a study of this text in early manuscript, see Andrew Glass, *Four Gāndhārī Saṃyuktāgama Sūtras: Senior Kharoṣṭhī Fragment 5*, with a contribution by Mark Allon. Gandhāran Buddhist Texts vol. 4 (Seattle: University of Washington Press, 2007).

Mahāvaipulya-mahāsaṃnipāta-Sūtra, *The Scripture of the Wise and the Foolish*,[70] the *Ten-stanza Vinaya*, and *Samantapāsādikā*. In *Samantapāsādikā*, it was listed with coconut together. Xuanzang has witnessed numerous *āmra* trees in India. However, our erudite master Daoxuan did not mention it in his list. Thus, it seems that at least in the North, it was not well known in the Chinese Buddhist community in Daoxuan's era or at least Daoxuan thought that it was not a significant fruit to his contemporary monastic community.

Furthermore, *ficus carica* (Skt. *udambara*, Ch. *wuhuaguo* 無花果, Jp. *ichijiku*, New Persian. *anjīr*) was also very common,[71] but was omitted in Daoxuan's list. Laufer argued that this fruit tree was imported from either Persia or India to China, not earlier than in the Tang Dynasty. In the Ming Dynasty, Li Shizhen 李時珍 (1513?–1593) said that this fruit tree was commonly cultivated in southern China, in particular, Jiangsu, Zhejiang, Hunan, Hubei, Fujian, and Guangdong; and especially in Guangdong, its transliterated name of Sanskrit, *youtanbo* 優曇缽, was used, while another transliteration in Chinese

For an English translation based on Pali edition, see Bodhi Bhikhu, *The Connected Discourses of the Buddha: A New Translation of the Saṃyutta Nikāya*, 2 vols. (Oxford: Pali Text Society, 2000). For a study on Sanskrit manuscript, see Ernst Waldschmidt, "Central Asian Sūtra Fragments and Their Relation to the Chinese Āgamas," in *Die Sprache der ältesten buddhistischen Überlieferung/The Language of the Earliest Buddhist Tradition*, ed. Heinz Bechert, Symposien zur Buddhismusforschung 2 (Göttingen: Vandenhoeck and Ruprecht, 1980), 136–74. Enomoto Fumio, "On the Formation of the Original Texts of the Chinese Āgamas," *Buddhist Studies Review* 3, no. 1 (1986): 19–30; Enomoto Fumio, "Sanskrit Fragments from the *Saṃyuktāgama* discovered in Bamiyan and Eastern Turkestan," in *Sanskrit-Texte aus dem buddhistische Kanon: Neuentdeckungen und Neueditionen*, erste Folge, *Sanskrit-Wörterbuch der buddhistischen Texte aus den Turfan-Funden*, Beiheft 2, eds. by Fumio Enomoto, Jens-Uwe Hartmann, and Hisashi Matsumura (Göttingen: Vandenhoeck & Ruprecht, 1989), 7–16; Enomoto Fumio, *A Comprehensive Study of the Chinese Saṃyuktāgama*, pt. 1, *Saṃgītinipāta* (Kyoto: Heirakuji shoten, 1994); Enomoto Fumio, "On a Sanskrit Fragment of the *Saṃyuktāgama* in the Hoernle Collection: The Chronology of the Chinese Version of the *Saṃyuktāgama*, Vasubandhu and Paramātha's Translation of the *Abhidharmakośabhāṣya*," in *Sakurabe Hajime koki kinen ronshū: Shoki Bukkyō kara abudaruma e* 初期佛教カらアビダルマへ (From Early Buddhism to Abhidharma), ed. Sakurabe Hajime (Kyoto: Heirakuji shoten, 2002), 139–53.

70 For the study of this text, see Victor H. Mair, "The Khotanese Antecedents of the *Sūtra of the Wise and the Foolish* (Xianyu jing)," in *Buddhism across Boundaries: Chinese Buddhism and the Western Regions*, ed. John R. McRae and Jan Nattier eds. (Taipei: Fo Guang Shan Foundation for Buddhist & Culture Education, 1999), 361–420; Victor H. Mair, "The Linguistic and Textual Antecedents of *The Sūtra of the Wise and the Foolish* (*Hsien-yü ching*) with an appended translation of 'Sudatta Raises a Monastery'," *Sino-Platonic Papers* 38, April 1993.

71 Maku Takamaro 満久崇麿, *Butten no shokubutsu jiten* 仏典の植物事典 (*Dictionary of Plants in Buddhist Texts*) (Tokyo: Yasaka Shobo, 1977), 30–32.

"*yingri*" might be from New Persian. Laufer also cited Xuanzang's record that this fruit was planted in India.[72] In Daoxuan's era, the Chinese Buddhist community was presumably still unfamiliar with this fruit; hence Daoxuan did not include it in his list.[73]

In Daoxuan's classification, the five grains (*wugu* 五穀) included the house grain (*fanggu* 房穀), the loose grain (*sangu* 散穀), the horn grain (*jiaogu* 角穀), the beard grain (*manggu* 芒穀),[74] and the cart grain (*yugu* 輿穀).[75] These names of the five grains originally appear in Chinese translations of Buddhist scriptures.[76] Yet the name "five grains" (*wugu*) also comes from indigenous Chinese tradition, and is attested as early as in the Zhou Dynasty (fifth century BC). The specific names of five grains used together do not appear anywhere except in Daoxuan's text discussed here. In Huilin's *Pronunciation and Meaning of All Scriptures* (*Yiqiejing yinyi*), citing from Yang Chengtian's 陽承天 dictionary *Assembly of Characters* (*Zitong* 字統), the names of five grains appear as *suigu, sangu, jiaogu, qigu,* and *shugu.*[77] Furthermore, according to

---

72 Berthold Laufer, *Sino-Iranica*, 411–14. Based on a note in Duan Chengshi's *Youyang zazu* 酉陽雜俎, Li Fan also suggests that it was imported to China in the Tang Dynasty. See, Li Fan, *Zhongguo zaipei zhiwu fazhanshi*, 221.

73 Heritaka (*helile*) is another plant that Daoxuan does not mention in his list of plants. It is a very important plant in Buddhist tradition, also frequently appears in Chinese botanical documents, such as *Tang bencao, Tujing bencao.* For a study about this plant in Tang China, see Li Hongbin 李鴻賓, "Dagu wenshu suojian bintie toushi zhuwu bianxi 大谷文書所見鑌鐵鍮石諸物辨析 (Studies on the Wrought Iron and Brass Mineral in the Turfan Manuscripts of the Otani Collections)," *Wenshi* 34 (1992): 148–51.

74 *miscanthus sinensis.*

75 In Tantric Buddhism, five grains included barley (*Hordeum vulgare*, Ch. *damai*, Skt. *yava*), wheat (*Triticum aestivum*, Ch. *xiaomai*, Skt. *godhūma*, paddy (*Oryza*, Ch. *daogu*, Skt. *sāli*), small bean (Ch. *xiaodou*, Skt. *masūra*), and oriental sesame (*Sesamum indicum*, Ch. *huma*, Skt. *atasī*).

76 Modern scholars have a different system to classify all grains. See Yamaguchi Hirofumi 山口裕文 and Kawase Makoto 河瀬眞琴 ed., *Zakkoku no shizenshi: sono kigen to bunka o motomete* 雑穀の自然史: その起源と文化を求めて (*The Natural History of Miscellaneous Grains: Looking for Their Origins and Culture*) (Sapporoshi: Hokkaidō Daigaku Tosho Kankōkai, 2003), 3–29. And five grains play important roles in medieval Chinese Buddhist ritual. For example, as Koichi Shinohara noted, "In the maṇḍala practice for Prajñāpāramitā, seven jewels and five grains are buried at the center of the maṇḍala at a depth of one hand (vitasti)." See his *Spells, Images, and Maṇḍalas: Tracing the Evolution of Esoteric Buddhist Rituals* (New York: Columbia University Press, 2014), 65.

77 Huilin, *Yiqiejing yinyi* 一切經音義 (ch. 16), *T.* no. 2128, 54, 403b. Modern scholarship rarely touches this issue. For example, Liang Jiamian traces the tradition of Chinese botanical taxonomy back to an ancient dictionary *Erya.* Liang Jiamian 梁家勉, "Zhongguo gudai zhiwu xingtaixue fenleixue de fazhan 中國古代植物形態學分類學的發展 (The Development of the Plant Forms and Classifications in Ancient China)," in *Liang Jiamian nongshi wenji* 梁家勉農史文集 (*Collected Papers by Liang Jiamian on Agricultural History*), ed. Ni Genjin 倪根金 (Beijing: Zhongguo nongye chubanshe, 2002), 413–23.

Daoxuan, these five grains have the following variant names: foxtail millet (*su* 粟), sorghum (*shu* 秫, or *gaoliang* 高粱), broomcorn millet (*shu* 黍, or *ji* 稷),[78] paddy (*dao* 稻),[79] wheat (*mai* 麥),[80] perilla (*ren* 荏),[81] all kinds of beans (*dou* 豆),[82] and even linseed (*ma* 麻).[83]

Interestingly, in Daoxuan's classification, both perilla and linseed are listed as grains. This idea certainly comes from indigenous Chinese tradition. Specifically, perilla was recorded in a botanical work titled *Illustrated Scripture of Herbs* (*Tujing bencao* 圖經本草) by Su Song in the Song Dynasty, in which it is said that Perilla was called "*su* 蘇" in Southern China and "*ren* 荏" in northern China.[84] It seems that Daoxuan used its northern name. Sesame was a popular food in early medieval China.[85] The horn grain comprises all kinds of beans and linseed (*jusheng* 巨勝).[86] Linseed was called "*jusheng*" in Daoxuan's list. This Chinese name also appeared in an ancient Chinese herbological work *Herbological Classic* (*bencaojing*); while in this work, linseed or "*jusheng*" was viewed as the same plant as oriental sesame or *huma* 胡麻 believed to be imported by Han general Zhang Qian, as recorded in the "Section about Western regions" ("*Xiyu zhuan* 西域傳") of *History of Han* (*Han shu* 漢書). However, later on many scholars made distinctions between "*jusheng*" and "*huma*." For instance, Tao Hongjing (456–536) says that the square-stem one is "*jusheng*"; while the round-stem one is "*huma*." See his commentary on the *Herbal Classic of the Divine Farmer* (*Shennong bencaojing* 神農本草經). Ge Hong suggests that "*jusheng*" is one kind of "*huma*," because it has two pods on one horn. In Su Song's *Illustrated Scripture of Herbs*, a separate category (ch. 18) called

78 Su Song, *Tujing bencao*, 538–39. Su Song says this grain is the head of five grains. In north China, people called it "dark cereal."

79 Su Song, *Tujing bencao*, 537–38. Also see Li Fang, *Zhongguo zaipei zhiwu fazhanshi*, 22–36.

80 Su Song, *Tujing bencao*, 532–33.

81 Nowadays it is well known as an oil plant.

82 Su Song, *Tujing bencao*, 534–36.

83 In Yuan Dynasty, planting beans and sesame together was recorded in a work titled *Zhongyi biyong* by Wu Yi, supplemented by Zhang Fu, edited and annotated by Hu Daojing (Beijing: Nongye chubanshe, 1962), 15. For a general survey of cultivated plants for food, see Li Fan, *Zhongguo zaipei zhiwu fazhanshi*, 22–66.

84 Su Song, *Tujing bencao*, 514–16.

85 Jia Yingxie, *Qimin yaoshu*, "huma," 13, ch. 2. Sesame was an essential food in ancient India, along with rice, butter, and salt in the Buddha's era; see Singh, *Planet, Plants and Animals*, 115.

86 Linseed's modern botanical name is *Linum usitatissimum*. For a detailed study on *huma*, also see Berthold Laufer, *Sino-Iranica*, 288–96. According to Laufer, this plant was cultivated mostly in north China. It has been discussed in Su Song's *Tujing bencao*, see Su Song, *Tujing bencao*, 526–27.

the "section of rice" (*mibu* 米部) is listed, including *huma*, two kinds of beans, millet, sorghum, grain, and so on.[87] This category includes more than the five grains listed by Daoxuan. Many of these grains, vegetables, and fruits were important for medieval Chinese Buddhist rituals. For example, as the *Collected Dhāraṇī Sutra* (Tuoluoni jijing 陀羅尼集經) indicated, in preparing for the All-gathering Maṇḍala Ceremony, the following stuff should be also prepared along with many other material offerings: huma, white honey, butter, pure human oil, white mustard seeds, pine oil butter, pomegranate, big chestnut, dried jujube, dried persimmon, pear, hutao, dried grape, etc.[88] Donating these fruits and plants would tremendously strengthen the ties between medieval Chinese Buddhist monastic community and the supportive lay community.

Though Daoxuan does not mention what sources he used for classifying these plants, it seems that his classifications benefited from various Vinaya traditions. Given that he was ordained in the *Four-part Vinaya* tradition, it is natural to conclude that his classification of economic plants was based on the *Four-part Vinaya* tradition. Yet in Daoxuan's text, some of names for plants are different from *Four-part Vinaya*, which he might have modified them for the sake of his Chinese readers. For example, in *the Dharmagupta-Vinaya*, the stem type (*jingzhong*) appears as the branch type (*zhizhong*),[89] while in *Great Assembly Vinaya* (*Mahāsanghika-vinaya* 摩訶僧祇律) it is exactly as Daoxuan termed it.[90]

Daoxuan's classification is different from traditional Chinese botanical taxonomy. In Tang times, some Chinese medicinal works also offer classifications. Tao Hongjing's work *Variorum of Herbological Classic* (*Bencao jing jizhu* 本草經集注) classifies plants into five categories: grasses, trees, fruits, grains, and vegetables, based on the principles of the forms and uses of the plants. Each category includes three classes (*sanpin*): high, medium, and low classes. In early Tang period, the Tang government sponsored a project to compile a work entitled *Newly Compiled Herbology* (*Xinxiu bencao*), which also followed this system.[91] So does Su Song's *Illustrated Scripture of Herbs* (*Tujing bencao*). Daoxuan's classification focuses on the main economic plants

87 Su Song, *Tujing bencao*, 526–39.

88 see *Tuoluoni ji jing*, juan 12, T. 18, no. 901, 894a1–3. For the detailed description on the ceremony, see Shinohara, *Spells, Images, and Maṇḍalas*, 65–88.

89 *Sifen lü*, *T.* no. 1428, 22, 641c.

90 *Mahāsanghika-vinaya* mentions five types of plants, with slight differences. For example, there is no *zazhong*; instead, it uses a name of "type of heart" ("*xinzhong*"). See *Mahāsanghika-vinaya*, *T.* no. 1425, 22, 339a-b.

91 Zhongguo zhiwu xuehui 中國植物學會 ed., *Zhongguo zhiwu xue shi* 中國植物學史 (*A History of Plant Science in China*) (Beijing: Kexue chubanshen, 1994), 41. Modern phytotaxonomy was only introduced to China in mid-nineteenth century (145).

used in monastic communities, which includes vegetables, grains, and fruits. He does classify some trees, but his classification seems to be limited to "trees bearing fruit."

Still being confined as a Buddhist master, Daoxuan does not follow traditional taxonomy and list grasses and trees in separate categories. Chinese traditional taxonomy of plants can be also found in the *Inner Classic of the Yellow Empire* (Huangdi neijing 黃帝內經), which listed five grains, five fruits, and five vegetables. In this Chinese medical text, five grains included rice, red bean, wheat, soybean, and corn. Five fruits included jujube, plum, apricot, chestnut, and peach. Five vegetables included mallow, leek, green onion, and agastache. In Chinese traditional medical thought, these five plants were connected with five flavors (sweet, sour, bitter, salty, and spicy) and also five organs (liver, heart, lung, spleen, and kidney) in human body. In the Sui dynasty, Xiao Ji (?–614) again mixed them with the theory of Five phases in his *Great Meaning of Five Phases* (*Wuxing dayi* 五行大義).[92] Daoxuan does not even mention the medicinal herbs that played a significant role in traditional botanical learning in medieval China. Daoxuan does not intend to work on plant taxonomy; his classification only serves to deal with plants of economic value to the Buddhist monastic communities.[93]

Conclusions

Although the Buddhist monastic community was originally designed to mainly serve as a spiritual community for the monastic members to concentrate on their spiritual cultivation for enlightenment, it should be supported by the lay community. The material foundation for a monastic community cannot be separated from natural world. Animals and plants were an indispensable part of the monastic community in medieval China. while handling animals and plants in the monastic community, Chinese monastic members, especially the masters in charge of the monastic administration, must develop their knowledge of animals and plants. In this sense, as one of

---

Mori Shikazō 森鹿三 studies a manuscript of *Xinxiu bencao* preserved in a Japanese private library. See his *Honzōgaku kenkyū* 本草學研究 (*A Study on the Herbology*) (Ōsaka: Takeda Kagaku Shinkō Zaidan, Kyōu Shorin, 1999), 2–34.

92 Xiao Ji, *Wuxing dayi*, juan 3, Nanjing: Jiangsu guji chubanshe, 1988, *Yuanwei bie cang*, vol. 70, 141–12; for a study, see Liu Guozhong 劉國忠, *Wuxing dayi yanjiu* 五行大義研究 (*A Study on the Great Meanings of Five Phases*) (Shenyang: Liaoning jiaoyu chubanshe, 1999), 210.

93 They can be also viewed as food plants; see Lambert Schmithausen, *The Problem of the Sentience of Plants in Earliest Buddhism* (Tokyo: The International Institute for Buddhist Studies, 1991), 36–46.

the top leaders of the monastic community. Daoxuan wrote a text for offering his suggestions about knowing and handling both animals and plants. His botanical knowledge not only came from the Buddhist Vinaya tradition rooted in comprehensive Buddhist literature for governing the moral and spiritual life of the Buddhist monastic members, but also benefited from his daily experience of living in various monastic communities in China. For Daoxuan, unlike Buddhas and Bodhisattvas, plants were not entirely foreign to medieval Chinese Buddhist community, though some plant species were indeed exotic. In the meantime, China was a vast empire at that time, so there were a lot of regional differences for growing and cultivating different plants, depending on natural environment and climate conditions. On the one hand, the Chinese Buddhist leaders such as Daoxuan inherited the early Buddhist traditions and acknowledged the material and cultural uses of some plants in Chinese Buddhist community. For example, the lotus flower still served as a sacred plant to play crucial symbolic role in medieval Chinese Buddhism. Chinese Buddhists indeed followed the rules to abstain from consuming five forbidden vegetables. On the other hand, medieval Chinese Buddhist masters attempted to come to terms with the differences between Buddhist plants and Chinese plants.

First issue is the difference between China and the Western Regions, especially South Asia. Daoxuan's knowledge of plants combines Buddhist tradition with Chinese indigenous tradition. He not only justifies his knowledge of plants based on his reading of early Buddhist monastic code that was rooted in natural environment of early South Asia, but also considers contemporary Chinese botanical knowledge. His broad knowledge on Buddhist texts offered him a unique position in discussing plants in Chinese monastic community and helped him jump from early Buddhist tradition to Chinese tradition. Although he was ordained and trained in the *Dharmagupta-Vinaya* tradition, he was also familiar with other Vinaya texts that have been translated into Chinese. He lived in Chang'an, one of the most cosmopolitan cities in the medieval world, so he was familiar with Xuanzang's record about the journey to the Western Regions. However, as the analysis above shows, Daoxuan's textual knowledge and practical experience with animals and plants in a certain historical context by no means could offer a precise image.

Second, there was a big difference for cultivating plants between northern region and southern region within the Chinese empire, due to different natural environments and climate conditions. Daoxuan's classification of plants is more influenced by Chinese botanical tradition than by Buddhist tradition, for he classifies plants into mainly three groups: grains, fruits, and trees. Fourth, in Daoxuan's list, it seems to be apparent that Daoxuan seemed to be more familiar with the plants that were cultivated in north China.

It indicates that Daoxuan's knowledge of these animals and plants seems to be based on his daily experience in the monastic community in north China. This experience was historical and empirical, different from his background and training in various textual traditions of monastic code. Ironically, his ideal model of Chinese Buddhism monasticism is the Buddhist monasticism in south China. Through an examination of Daoxuan's ideas about dealing with plants, we can also conclude that the ownership of plants was restricted to Buddhist community, and prohibited to individuals.

Finally, this study also attempts to emphasize the importance of studying plants in medieval Chinese religions. Contemporary scholarship of Chinese Buddhism has focused on Buddhist philosophical concepts, the representations of Buddhist saints, eminent monks, rituals, socio-economic issues, and state-church relations. However, this study suggests that Buddhist botanical knowledge was also very important for understanding the spread of Buddhism in medieval Chinese society, for many Buddhist concepts in daily life such as the classification of plants also experienced some crucial transformations from early Buddhist culture to Chinese culture. This botanical transformation, similar to the rise of the ghost festival, was a crucial part of the process that Buddhism firmly took root into medieval Chinese society. Furthermore, while Daoxuan's text was eventually incorporated into the Buddhist canon and widely acknowledged by the Buddhist community across the Chinese empire in the later period, it completed the canonization of the Buddhist botanical knowledge in China.

Chapter 3

ANIMAL DIVINATION AND CLIMATE: AN ENVIRONMENTAL PERSPECTIVE ON THE CULT OF THE PIG

Introduction

Although previous two chapters focused on plants, animals were not less significant in shaping the material and spiritual life of ancient China. In the classical era, the tradition of observing nature was developed in the context of the *Yin-Yang* and the Five-phase theories, which were later incorporated into the ancient arts of divination, including the technique of predicting weather changes by observing the behavior and health of animals. The observation of the close connection between animals and weather developed into the worship of animals, that is, which led to the rise of the cult of animals. This chapter aims to analyze the rise of the cult of the pig in ancient China, with special reference to the context of environmental and climate change.

As one of the most common species among domesticated animals in traditional China's agricultural civilization, the pig occupies a unique position in both the material and the spiritual life of the Chinese people, which some comprehensive Western-language works on pigs in the past decade did not fully acknowledged.[1] In the Central Plain of the Chinese empire, where the so-called Yellow River civilization had its origin, the pig was the most important source of meat consumed in daily life, thus contrasting sharply with the civilizations of the Jewish, Muslim, and Christian people who lived in the ancient Near East and the Middle East where pork was a food taboo. Contemporary scholars have debated about

---

1 Bret Mizelle, *Pig* (London: Reaktion Books, 2011); Mark Essig, *Lesser Beasts: A Snout-to-Tail History of the Humble Pig* (New York: Basic Books, 2015); Richard Lutwyche, *The Pig: A Natural History* (Princeton: Princeton University Press, 2019).

the origins of the view of the people in these regions that pigs are unclean, and the taboo against pork in these areas.[2]

One of the most notable perspectives on this issue is Marvin Harris' ecological and environmental theory. For Harris, pigs are hated not only because they are perceived as unhygienic, given that they frequently roll around in their own feces and urine, but also because they compete with humans for the same basic nutrients, such as those found in grain. What interests us is Harris' observation that pigs have to roll around in their own feces and urine in order to cool off. Harris notes that the behavior of pigs in the Middle East may be influenced by the local climate and the environment. Based on his understanding of animal physiology, Harris suggests that, due to the lack of protective hair and the inability to sweat, the pig must dampen its skin with external moisture by wallowing in fresh clean mud."[3] This observation reminds us of sources in ancient China that describe such behavior in pigs.

Interestingly, in ancient Chinese sources, the pig was portrayed in the earliest texts as a beast living near water. Later, it became symbolic of rain, and its behavior was used to predict rainstorms. Inspired by Harris' hypothesis and discussion, this chapter examines these Chinese sources in an attempt to understand how Chinese writers observed and portrayed the pig from ecological and environmental perspectives. The aim is to analyze the sophisticated relations among animal behavior, ecology, and environment, as well as religion and the associated rituals. Because the behavior of pigs is also influenced by the way in which they are raised, we are also interested in investigating whether the change from a free, foraging way of life to confinement in pens may have had an impact on observers' understanding of pig behavior and associated beliefs and practices.

Contemporary scholarship has focused primarily on textual studies and folklore surrounding the pig as a god of water and a god of thunder. Chinese and Japanese scholars have concentrated on deities that often appear as pigs

---

2 Marvin Harris, *Cows, Pigs, Wars, and Witches: The Riddles of Culture* (New York: Random House, 1974), 35–57; Paul Diener et al., "Ecology, Evolution, and the Search for Cultural Origins: The Question of Islamic Pig Prohibition [and Comments and Reply]," *Current Anthropology* 19, no. 3 (1978): 493–540; Frederick J. Simmons, "Traditional Use and Avoidance of Foods of Animal Origin: A Culture Historical View," *BioScience* 28, no. 3 (1978): 178–84; Brian Hesse, "Pig Lovers and Pig Haters: Patterns of Palestinian Port Production," *Journal of Ethnobiology* 10, no. 2 (1990): 195–225; Richard A. Lobban Jr., "Pigs and Their Prohibition," *International Journal of Middle East Studies* 26, no. 1 (1994): 57–75; Essig, *Lesser Beasts*, 43–52, 89–103; Lutwyche, *The Pig*, 148–51.

3 Harris, *Cows, Pigs, Wars, and Witches*, 42–43.

in traditional literature, folklore, and drama, such as the god of water (*shuishen* 水神), the god of thunder (*leishen* 雷神), and the god of the river (*hebo* 河伯), as well as the Marshall of the Celestial Canopy (*Tianpeng yuanshai*, or *Zhu Bajie*). For example, Min Xiangpeng's very short essay traces the changing images of Zhubajie from the perspective of textual studies, while he does draw attention to the climate issue.[4] Such changes from one identity to another must be examined carefully in the broader philosophical and intellectual context of ancient China. Mark Elvin and Robert B. Marks' works remind us how environmental perspective could offer us new insights in understanding Chinese history.[5] This study will expand and revise Min's brief observations. It aims to add a fresh and concise document of demonstrating the sophisticated relationship between the animal cult and the environmental context in premodern China to the Western-language literature.

In addition, animal studies constitute a new trend that has developed during the past two decades, providing new methodologies for analyzing the relations between humans and animals, and encouraging us to take the life experiences of humans and animals into account in order to understand animal cults and their transformations. Some scholars have noted the interconnectivity between humans and animals in ancient China and argued that human morality could transform the animal realm from the perspective of early Chinese philosophical cosmology.[6] However, in documenting the changing images of the pig as a case study, this chapter attempts to demonstrate that animals and humans both lived in a changing environment, and their behaviors in relation to each other also changed to accommodate to changes in environmental conditions. Humans observed the behavior of animals and created images of animals for social and cultural, as well as religious purposes based on their behaviors. This chapter documents the processes by which these deities developed and provides a fresh perspective on the ecological and environmental context in which the pig's behavior was understood, depicted, and interpreted in the historical, cultural, and religious life of ancient Chinese people.

---

4 Min Xiangpeng, "Gudai qixiang guance, minjian Xinyang yu wenxue chuangzuo de duoxueke shijiao: Xiyouji Zhong Zhubajie shuishen xingxiang de yanbian," *Beijing shehui kexue*, no. 10 (2014): 67–72.

5 Mark Elvin, *The Retreat of the Elephants: An Environmental History of China*, New Haven: Yale University Press, 2004; Robert B. Marks, *China: Its Environment and History*, Lanham: Lexington Books, 2011.

6 Roel Sterckx, *The Animals and the Daemon in Early China* (Albany: State University of New York Press, 2002), 162; Deborah Cao, *Animals in China: Law and Society* (London: Palgrave Macmillan, 2015), 16.

Climate, Environment, Animal Behavior, and Husbandry

In ancient China, correlative thinking often led people to connect astronomical phenomena with social changes and animal behavior on the ground. People predicted rainfall by observing animal behavior, along with the movements of the wind and the clouds. The earliest record of observing pigs' behavior can be found in the poem "Jagged Are the Rocks" ("*Jianjian zhi shi*" 漸漸之石) in the *Minor Odes* (Xiao Ya 小雅) section of the *Book of Poetry*.[7] The image of the pig dashing into the water together with the reference to the Hyades has the effect of placing the pig among astronomical phenomena. In the Tang Dynasty, Kong Yingda's 孔穎達 (574–648) textual research suggests that hogs running into the water and the moon is caught in the Net are signs that it will rain.[8] After that, many scholars through the ages agree with this explanation. Until the Qing Dynasty, Xu Ding 徐鼎 (1718–?) quotes the saying, "the hog likes the rain" in *Han's Explication of Poetry* (*Han shi shuo* 韓詩說).[9] Therefore, the ancient Chinese regarded the pig could predict rainfall, which seems a big issue that could not be found in other pig cultures, such as those in eastern Mediterranean region where the pig was regarded as an unclean animal because of its habit of wallowing in mud. Given the difference between the pig as a water animal in ancient China and the pig as a muddy animal in ancient Mediterranean region, this chapter attempts to present a new perspective that the different climate environment in East and West Asia motivated the behaviors of pigs. In ancient China, pigs were accustomed to the rainy, humid, and hot climate during the monsoon season, which was different from the dry and hot climate of West Asia.

Strictly speaking, it is not the pigs that predict rain but the humans, who predict it by observing the pigs' behavior. Pigs seem to sense that rain is coming and begin their wallowing behavior even before it has started. The humans use this observation to predict rain. As this section will show, there are solid scientific reasons for this behavior as well. When it is about to rain, the air becomes humid, and the pressure is low. This makes the pigs feel very hot. They are likely to run into rivers or muddy areas and swim or roll around in the mud to get cool. This association between pigs and impending rainfall makes it appear that the pigs have special knowledge or special powers of

---

7 Zhu Xi 朱熹, *Shiji zhuan* 詩集傳 (Commentaries on the Collection of Poems), compiled and annotated by Wang Yi 王逸 and Hong Xingzu 洪興祖 (Changsha: Yuelu shushe, 1989), 199; Arthur Waley trans., *Book of Songs* (London: Routledge, 2005, reprinted edition), 120.

8 Kong Yinda 孔穎達, *Mao shi Zhengyi* 毛詩正義, compiled by Ma Xinmin 馬辛民 and Li Xueqin 李學勤 (Beijing: Beijing daxue chubanshe, 2000), 15.

9 Xu Ding 徐鼎, *Maoshi mingwu tushuo* 毛詩名物圖說 (Illustrated Description of Animals and Plants in Book of Songs), compiled by Wang Chenglüe 王承略 (Beijing: Tsinghua daxue chubanshe, 2006), 199.

perception, beyond that of human beings, which could easily be seen as having mysterious or magical qualities.

From a biological perspective, pigs need a lot of water to survive. First of all, water is the most critical component of a pig's body. It constitutes 80% of a newborn pig's body, and it is vital for the pig to maintain this percentage of water. If it is even slightly higher or slightly lower, the pig may become ill. A loss of 10–20% of its total body water will lead to death.[10] Second, for both domestic pigs and wild boars, their thick porcine subcutaneous fat layer, combined with their lack of functional sweat glands, necessitates wallowing in mud to prevent hyperthermia.[11] M. B. M. Bracke has shown that as many as forty-eight academic papers contain citations about the wallowing behavior of pigs and wild boars, and twelve papers contain citations about wallowing in related species. The common perception is that pigs wallow mainly to cool down, to protect themselves from sunburn, and to remove ectoparasites. However, there is little scientific evidence for any reason other than thermoregulation.[12] There is a consensus among academics that the prime function of wallowing is to prevent hyperthermia.[13]

The same behavior in pigs has been observed in different parts of the world. In ancient times, pigs live in a wide range of habitats, but these habitats always have sources of water nearby. These sources allow pigs to perform what seems to be their favorite activity: wallowing. References to the wallowing of pigs in the mud also occur in early Western literature. For example, the Roman poet Horace (65–8 BCE) noted the muddy hog,[14] though his poem seems to use the so-called hog delighting in mire as a metaphor for a way of life. Plutarch (AD 46–119) also noted that, when some young people came to Athens to visit Cebes, a disciple of Socrates, "As they were walking along the street of the statuaries past the law-courts, they were met by a drove of swine, covered with mud."[15] The connection between swine and mud in ancient Europe and the Middle East seems more common,[16] in contrast to the connection

10 Palmer J. Holden and M. E. Ensminger. *Swine Science* (Princeton: Prentice Hall, 7th edition, 2005), 296–301.

11 Bian Lianquan 邊連全 ed., *Zhu siyang jishu* 豬飼養技術 (Pig Feeding Technology) (Shenyang: Dongbei daxue chubanshe, 2010), 10.

12 Marc B. M. Bracke, "Review of Wallowing in Pigs: Description of the Behaviour and Its Motivational Basis," *Applied Animal Behaviour Science* 132, no. 1–2 (2010): 1–13.

13 D. L. Ingram, "Evaporative Cooling in Pig," *Nature* 207 (1965): 415–16.

14 Horace, *The Works of Horace*, Epistle II. trans. by C. Smart, revised by Theodore Alois Buckley (London: George Bell & Sons, 1888), 233.

15 Plutarch, *Plutarch's Moralia*, vol. 7 (London: W. Heinemann, 1927), 407.

16 See Marvin Harris's note on the pig's behavior of wallowing in mud in the Middle East; see Harris, *Cows, Pigs, Wars, and Witches*, 36; Salima Ikram noted the same thing in ancient Egypt; see Salima Ikram, *Choice Cuts: Meat Production in Ancient Egypt* (Leuven: Peeters, 1995), 33.

between swine and water in ancient China, though the wallowing behavior of pigs seems to be a common phenomenon in the ancient writings of both China and Europe.

What seems to be unique to China, however, is that the idea that the pig wallows immediately before it rains contributed to the rise of the cult of the pig in which the pig was portrayed as the controller of rainfall. China's unique climate may be a key to understanding the ancient philosophy and religion related to this issue. Compared with Europe and elsewhere in the world, East Asia has a monsoon climate, where the hot summers are accompanied by high rainfall (see Charts 3.1–3.3).[17] The early Egyptian, Roman, and Greek civilizations in Europe and North Africa had for the most part a Mediterranean climate, with low rainfall in the summer and high rainfall

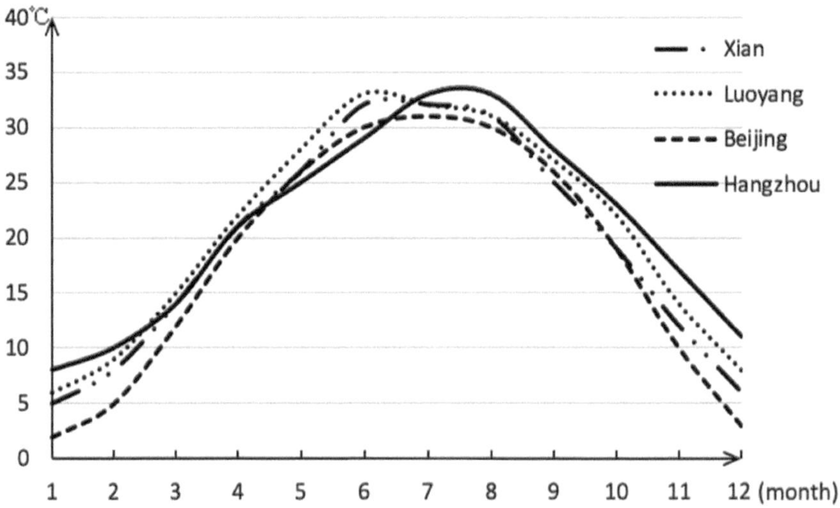

Chart 3.1: The annual highest temperature in some capital cities of premodern China.

17 Studies can be found in Gergana Yancheva et al., "Influence of the Intertropical Convergence Zone on the East Asian Monsoon," *Nature* 445 (2007): 74–77; Zhu Kezhen 竺可楨, "Zhongguo jin wuqiannian lai qihou bianqian de chubu yanjiu 中國近五千年來氣候變遷的初步研究 (A Preliminary Study on the Climate Change in China in nearly Five Thousand Years)," *Kaogu xuebao* 考古學報 (*Archaeological Journal*) 1 (1972): 15–38; Ge Quansheng et al., "Sui Tang shqi dongzhongbu diqu wendu bianhua de chongjian 隋唐時期東中部地區溫度變化的重建 (The Reconstruction of the Changing Temperatures in the Eastern and Central Regions of the Sui and Tang Periods)," *Kexue tongbao* 科學通報 (*Science Bulletin*) 55, no. 31 (2010): 3048–52; Michael E. Mann et al., "Proxy-Based Reconstructions of Hemispheric and Global

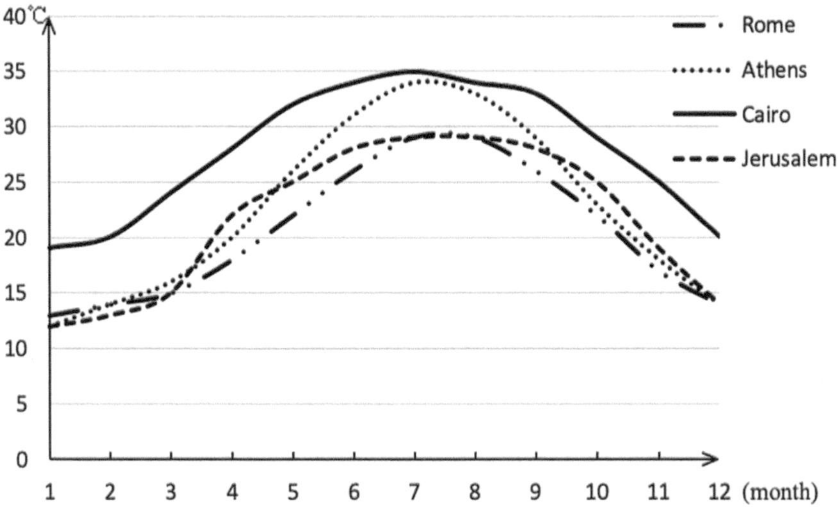

Chart 3.2: The annual highest temperature of some cities in eastern Mediterranean region.

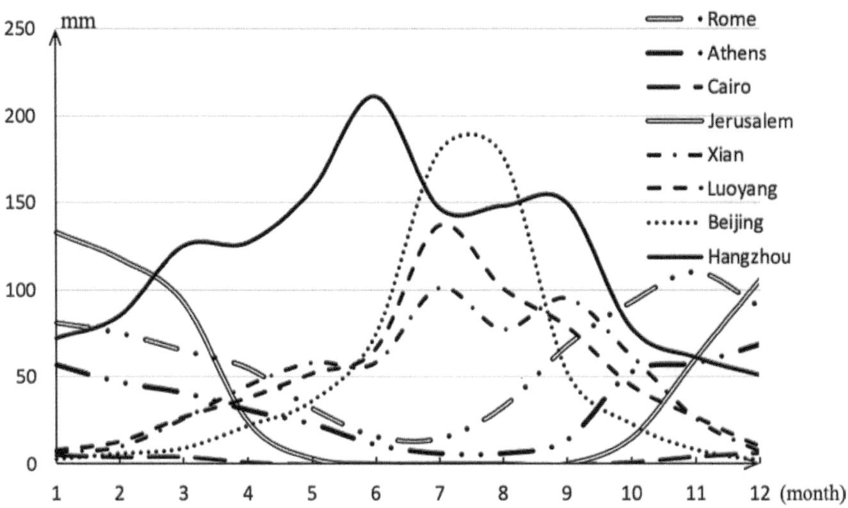

Chart 3.3: A comparison of historical rainfall lines between Chinese ancient cities and Mediterranean cities.

Surface Temperature Variations over the Past Two Millennia," *Proceedings of the National Academy of Sciences* 105, no. 36 (2008): 13252–57; Anders Moberg et al., "Highly Variable Northern Hemisphere Temperatures Reconstructed from Low and High Resolution Proxy Data," *Nature*, no. 433 (2005): 613–17; Yang Bao et al., "General Characteristics of Temperature Variation in China during the Last Two Millennia," *Geophysical Research Letters* 29, no. 9 (2002): 38.

in the winter (Chart 3.3). Thus, the hottest weather in these regions had (and still has) the lowest rainfall. This does not explain why pigs were not worshiped in these areas, but the contrast does help to highlight the contributing factor of climate and rainfall in the development of pig worship in China.

Furthermore, hogs were not domesticated in ancient China; they lived in the wilderness. Thus, it was easier for people to observe these hogs moving toward the river or water. Even for domesticated pigs, the practice of rearing pigs in ancient China experienced a shift from natural husbandry to ranching husbandry during the eleventh century.[18] After the Song Dynasty, particularly in the eleventh century, the practice of rearing pigs in ancient China experienced a shift from natural husbandry to ranching husbandry, which means that the pigs might be no longer observed to approach the river in the wilderness when the rainstorm was coming up. Many scholars in the Song Dynasty noted that the pigs were reared near the river with rich grass, so the pigs often had white feet because they washed their feet in the nearby water.[19] Yet later in the Ming Dynasty, the prediction of rainfall associated with the pigs was not acknowledged by some scholars,[20] which was due to the ranching husbandry of pigs. In sum, under conditions of natural husbandry, pigs can easily access water areas or rivers in the wilderness, thus enabling people to observe that they move toward water when it is about to rain. Given the frequent rainstorms during the monsoon season in China, the pig's behavior of accustoming to run to water became more visible to local observers. Once pigs are reared under ranching husbandry, they no longer have free access to such water areas.

From Water Beasts to the Flood Gods/Deities

In ancient times, people all over the world observed changes in natural phenomena and in the behavior of animals. The term "phenology" was coined in mid nineteenth-century Europe for the observations of natural

---

18 Many scholars briefly mentioned the husbandry of pig and meat business in ancient and modern China, but they do not pay much attention to the transformation of rearing pigs; see Elvin, *The Retreat of the Elephants*, 214 and 463; Marks, *China: Its Environment and History*, 293; Mindi Schneider, "Reforming the Humble Pig: Pigs, Pork, and Contemporary China," In *Animals through Chinese History: Earliest Times to 1911*, eds. by Roel Sterckx, Martina Siebert, and Dagmar Schäfer (Cambridge: Cambridge University Press, 2018), 233–43.

19 Luo Yuan 羅願, *Erya yi* 爾雅翼 (Ji'nan: Qilu shushe, 1995), 554; Zhang Zai 張載, *Shiji zhuan* 詩集傳 (Taipei: Taiwan shangwu yinshuguan, 1981), 175.

20 Zhang Cizhong 張次仲, *Daixuan shiji* 待軒詩記 (Shanghai: Shanghai guji chubanshe, 2003), 493.

phenomena such as climate and their correlations with animal behavior.[21] Although the term was not coined until the eighteenth century, the phenomenon existed long before. Theophrastus, for example, indicated that in ancient Greece, people practiced zoomancy by observing the features and behavior of animals. China also has a long tradition of phenology (*wuhou* 物候).[22] For instance, in the *Lesser Annuary of Xia* (*Xia xiao zheng* 夏小正, around the seventh to fourth centuries BCE), a farmer's calendar recorded numerous comments on the weather, the stars, and animal life for helping farmers make decisions on their agricultural schedule.[23] Other texts such as the *Records of Grand Historian* (*Shiji* 史記)[24] also recorded how ancient people practiced the technique of divination for making predications based on their observing changes in natural phenomena.[25]

Observing the behavior of pigs was part of this practice in ancient China. Two of the most common observations of pigs that appeared in traditional Chinese sources were the "bathing pig" (*yu zhu* 浴豬) and the "black pig crossing the river" (*hei zhu guo he* 黑豬過河).[26] The observations of these two phenomena led people in ancient China to associate the pig with water, rainfall, and rivers, as well as thunder. Eventually, the pig was deified. It is suggested here that Chinese observations of the pig's behavior in the context of the phenology and divination show that the pig's connection with water was transformed from a physiological phenomenon to astrological and astronomical phenomena, thus eventually the pig was transformed from a species of animal into a god.

Unlike in the Near East, where the ancient people concentrated on the association of pigs with rolling around in the mud, Chinese people saw the pig as a beast of water. Even as early as the third century BCE, ancient

---

21 Gaston R. Demarée and This Rutishauser, "From 'Periodical Observations' to 'Anthochronology' and 'Phenology': The Scientific Debate between Adolphe Quetelet and Charles Morren on the Origin of the Word 'Phenology'," *International Journal of Biometeorology* 55, no. 6 (2011): 753–61.

22 Zhu Kezhen and Wan Minwei 宛敏渭, *Wuhou xue* 物候學 (Phenology) (Expanded edition 增訂本, Beijing: Kexue chubanshe, 1973), 1.

23 Ho Peng Yoke, *Li, Qi and Shu: An Introduction to Science and Civilization in China* (Hong Kong: Hong Kong University Press, 1985), 121.

24 Sima Qian, *Shiji* 史記 (Beijing: Zhonghua shuju, 1982), 1305; As Zhang Shoujie's commentary on the *Shiji*'s section on 24 constellations, one of stars was called *kui* constellation, which refers to celestial pig who was in charge of water ditches and channels.

25 Min, "Gudai qixiang guance," 67–72.

26 Huang, *Xiangyu shu*, 4; Yang Liangcai and Dong Sen, *Yan hai (Sea of Proverbs)* (Lanzhou: Gansu shaonian ertong chubanshe, 1991), 485.

people in China had already attempted to predict rain, not only by noting changes in the winds and the clouds, but also by observing abnormalities in the behavior of pigs, as well as other animals. The *Book of Changes* (*Yijing* 易經) records the following prognostication: "When one sees a hog covered with mud on its back, … if one encounters rain as one goes, there will be good fortune 見豕負塗, 往遇雨則吉."[27] According to the explanation of *Shuogua* (说卦), the eight trigrams are said to correspond to eight animals, and the trigram of the Perilous Pit (*kan* 坎) corresponds to the hog (*shi*). The trigram of the Perilous Pit also corresponds to water, as one of the five phases, and is associated with rivers and streams.

According to the scientific study, temperatures in China were above average in the Han Dynasty "with two peaks around AD 50 and in AD 100–240. The peak at about AD 200 represents the warmest stage of the last two millennia, temperature was even higher than during the 20th century. The area that experienced warm conditions extended from Eastern to North Western China, the northern boundary of the farming zone even extended to the northern limit of the present farming-grazing zone. Following this period, temperatures decreased below average until AD 600."[28] The warm weather might have prompted pigs to the water for cooling. In the Han Dynasty, while commenting on the poem "Jagged Are the Rocks," Zheng Xuan 鄭玄 (127–200) noted that pigs have a habit of enjoying water,[29] rather than mud as we saw in the Middle East. The idea of the pig as a "water beast" (*shuichu* 水畜) seems to begin with the commentator Xiao Ji 蕭吉 (530–614) in his book, *The Great Meaning of the Five Phases* (*Wuxing dayi* 五行大義).[30] The discourse associating the *kan* trigram with water and the pig was well known to many

27 *The Book of Changes* might be compiled in the Zhou dynasty, but the part we deal with here might come from the third century BCE. Wen Yiduo's "Zhouyi yizheng leizuan" 周易義證類纂 (Exegetic Interpretation of the Book of Changes); see Wen Yiduo 聞一多, Wen *Yiduo quanji* 聞一多全集 (The Collected Works of Wen Yiduo) (Beijing: Sanlian shudian, 1982), 48.

28 Yang et al., "General characteristics of temperature variation in China during the last two millennia," *Geophysical Research Letters* 29, no. 9 (2002): 38.

29 Zheng Xuan 鄭玄, *Maoshi Zhengyi* 毛詩正義 (Beijing daxue chubanshe, 1999), 943.

30 Xiao Ji 蕭吉, *Wuxing dayi* 五行大義 (Essence of the Five Phases), edited by Qian Hang 錢杭 (Shanghai: Shanghai shudian chubanshe, 2001), 65; Liu Guozhong 劉國忠, *Wuxing dayi yanjiu* 五行大義研究 (A study of the General Principles of the Five Phases) (Shenyang: Liaoning jiaoyu chubanshe, 1999); Donald Harper, "The Textual Form of Knowledge: Occult Miscellanies in Ancient and Medieval Chinese Manuscripts, Fourth Century B.C. to Tenth Century A.D," in *Looking at It from Asia: The Processes that Shaped the Sources of History of Science*, ed. by Florence Bretelle-Establet (Dordrecht: Springer, 2010), 69–71.

classicists, such as Lu Dian 陆佃 (1042–1102), a scholar of the Song Dynasty, who said, "*kan* trigram represents pigs, and the *kan* trigram also contains a downward trend. Pigs can bow to their heads and like the wet environment. They are also 'water beasts'."[31] Hao Yixing 郝懿行 (1757–1825) continued this discourse and said, "The swine is an omen of floods because *kan* (the trigram) stands for both hogs and water."[32] Therefore, this tradition involving pigs with water, floods, and wet environment has a long history in the classical literature of premodern China.

Later on, in the Tang Dynasty, the pig became associated with the natural phenomenon of rainfall, and if a pig was seen to be bathing, this behavior was used to predict rain. This predication came from the observation of some phenomena in the sky and their connections with the pig's behavior on the ground. In his *Book on Observing Rain* (*Xiang yu shu* 相雨書),[33] the Tang scholar Huang Zifa 黄子發 says, "There are no clouds around the Big Dipper, but there are some in the Milky Way. If three puffs of cloud are linked together in the shape of a bathing pig, there will be heavy rain after three days."[34] This passage merges the idea of the bathing pig on the ground with the idea of black clouds in the sky. The former came from a long textual tradition that even could be traced back to the era of the *Book of Poetry*, and the latter was a daily experience of observing the sky in the medieval period.

In addition to bathing pigs, another observation involving these animals, this time a meteorological one, was used to predict rain. It is called "black pigs crossing the river." The idea of a "black pig crossing the river" refers to the appearance of certain types of clouds at night, which herald rain. According to the *Accounts of Marvels* (*Shu yi ji* 述異記), when the sky is very foggy at midnight, this is called the black pig crossing the river, and is a sign of rain.[35] The river being referred to is the Celestial River (*Tianhe* 天河), which refers to the Milky Way. This belief is credible because billowy or fragmentary clouds sometimes drift across the sky, acting as vanguards for

31 Lu Dian 陸佃, *Biya*, annotated by Wang Minhong (Hangzhou: Zhejiang daxue chubanshe, 2008), 44.

32 Ma Changyi 馬昌儀 ed. *Guben Shanhaijing tushuo* 古本山海經圖說 (Illustration of Ancient Classic of Mountains and Rivers) (Ji'nan: Shandong huabao chubanshe, 2001), 308.

33 Jeffrey P. Snyder-Reinke, *Dry Spells: Rainmaking, Power, and the State in Late Imperial China* (Cambridge, MA: Harvard University Asia Center, 2006), 122.

34 Huang, *Xiangyushu*, 4; Liu Zhaomin 劉昭民 ed., *Zhonghua qixiang xue shi* 中華氣象學史 (The History of Chinese Meteorology) (Taipei: Shangwu yinshu guan, 1970), 91–100.

35 Ren Fang 任昉, *Lieyizhuan deng wuzhong* 列異傳等五種 (Biographies of Anomalies and Four other Books) (Beijing: Wenhua yishu chubanshe, 1988), 460–508.

an ominous storm cloud that follows soon after, and it is possible to predict rain by observing the shape of such clouds at night.[36]

Even though in the Ming Dynasty, the pigs were often reared in the ranches rather than in the wilderness, observing the approach of the pigs to the water of river might not be part of daily experience at that time, the textual tradition continued in some agricultural works. The "black pig crossing the river" was a proverb quoted frequently in a number of mid-Ming works, such as *Monthly Climate and Phenology* (*Yue ling guang yi* 月令廣義), *Records of Mount Tianzhong* (*Tianzhong ji* 天中記, ca. 1569),[37] and *The Complete Compendium on Agricultural Administration* (*Nongzheng quanshu* 農政全書). The Ming scholar Yang Shen 杨慎 also said, in his *Danqian yulu* 丹鉛餘錄 (ca. 1550), "when there are black clouds in the Celestial River, we call this 'the black pig crossing the river,' and this implies rainfall."[38] Similar sayings are widespread in Chinese folklore, such as "the dark cloud passes through the Celestial River," that is, the Milky Way. It is also said that "a black pig crossing the river indicates a downpour 黑豬過河, 大雨滂沱,"[39] "if black pigs cross the stream, there will be rain 烏豬過溪要下雨,"[40] and "if the pig crosses the river, there will be heavy rain the next morning 豬渡河, 明朝雨多."[41] These proverbs are used by farmers in some areas even today to predict rain. Because both bathing pigs and "the black pig crossing the river" were used to predict rain, the ancients often linked them together without understanding why they were connected.

The pig is not only auspicious for bringing rain, but its appearance can also be an ominous sign of floods. Moreover, sometimes it is depicted visually as a cruel beast. Several early sources make an association between pigs and floods. This association of pigs with floods depicted pigs as powerful beasts who held destructive force that could disrupt daily life. This depiction also has a textual tradition. The *Classic of Mountains and Seas* (*Shanhai jing* 山海經, first century BCE) contains multiple references to the presence of wild pigs indicating an impending flood. In its section on the "Southern Mountains"

36 Min Xiangpeng, "Cong Pei Xingjian yuce baoyu tanqi 從裴行儉預測暴雨談起 (On Pei Xingjian Forecasting a Rainstorm)," *Zhongguo jianzai* 中國減災 (Natural Disaster Reduction in China), No. 8 (2009): 43.

37 Christoph Kaderas, *Die Leishu der imperialen Bibliothek des Kaisers Qianlong (reg. 1736–1796): Untersuchungen zur chinesischen Enzyklopädie* (Wiebaden: Otto Harrassowitz Verlag, 1998), 204–5.

38 Yang Shen 楊慎, *Danqian yulu* 丹鉛餘錄 (Records of an Exiled Man) (Shanghai: Shanghai guji chubanshe, 1992).

39 Yang and Dong, *Yanhai*, 485.

40 Yang and Dong, *Yanhai*, 109.

41 Yang and Dong, *Yanhai*, 538.

(Nanshan jing 南山經), it says: "The Changyou swine (*changyou zhi* 长右彘) looks like a monkey with four ears. It is named Changyou and makes a singing sound. A flood will certainly occur in the district or county (*jun xian* 郡縣) where it appears."[42] In his *Illustrated Eulogy to the Classic of Mountains and Seas* (*Shanhaijing tuzan* 山海經圖讚), the early medieval scholar Guo Pu 郭璞 (276–324) explains that the Changyou swine signifies that there will be a flood; when this beast appears, the flood will spread widely.[43] The section on the "Eastern Mountains" (*Dongshan jing* 東山經) also claims that the beast looks like Kuafu 夸父, a legendary figure in Han mythology. This beast is covered with swine hair on its body and sounds as if it is shouting. When it appears, a flood will wash the whole world.[44] This section also describes another beast, called Heyu 合窳, which has the body of a swine, a human face, a yellow body, and a red tail. In his *Illustrated Eulogy*, Guo Pu identifies the Changyou Swine with this beast called Heyu, emphasizing its greedy and cruel nature. It is able to swallow its prey without chewing. Interestingly, he connects this beast with the quality of extreme *Yin* 陰, because it brings rain.[45] Guo Pu explicitly refers to the theory of Yin and Yang of traditional Chinese philosophy to interpret the connection between pigs and floods. Since floods are dangerous and devastating phenomena and are associated with the *Yin*, the pig is also associated with *Yin*, and is depicted negatively as a greedy, cruel beast that brings floods. The association of pigs as cruel beasts that bring about dangerous flood disasters, which developed in early medieval China, continues in the literature of the later periods. For instance, a comprehensive collection called *Names of Things on Memory Beads* (*Shiwu ganzhu* 事物紺珠), compiled by Huang Yizheng in 1591, notes that the beast Heyu looks like a pig with a human face, and consumes blood as food.

From Earthly River Uncle to Heavenly River God

As noted in the previous section, the appearance of the pig is an ambivalent sign, signifying both beneficial rainfall and disastrous flood. The pig was also transformed from a water beast to a god of water, called the

42 Yuan Ke 袁珂 ed. and annotated, *Shanhaijing jiaozhu* 山海經校注 (The Classic of Mountains and Seas with Punctuations and Annotations) (Shanghai: Shanghai guji chubanshe, 1980), 31.
43 Zhang Zongxiang 張宗祥 ed., *Zuben Shanhaijing tuzan* 足本山海經圖贊 (Full-Length Eulogistic Picture for the Classic of Mountains and Seas), annotated edition of Guo Pu's 郭璞 *Shanhaijing tuzan* (Beijing: Gudian wenxian chubanshe, 1958), 3.
44 Zhang, *Zuben Shanhaijing tuzan*, 92.
45 Zhang, *Zuben Shanhaijing tuzan*, 21.

"River Uncle" in early medieval Chinese sources. As we will demonstrate in this section, the terrestrial River Uncle gradually became the Celestial River God, given the correlative thinking centered on the Yin-Yang and Five-phase theories that deeply rooted in traditional Chinese thought. Writers seem to have constructed different images of the pig by drawing upon multiple sources both from their ancient social and cultural context and from their spiritual life. Elements taken from traditional philosophy, literature, history, and religion combined to form changing images of the pig, as both a beast and a god, associated with water, rivers, and floods.

As they observed such natural phenomena as pigs bathing in water and black pigs crossing the river, ancient people associated pigs with water and also used their behavior to predict rainfall. They gradually came to believe that pigs had supernatural qualities. It turns out that pigs are identified with specific popular deities such as the River God or River Uncle (hebo 河伯), the Water Spirit (*shui shen* 水神), and the rain god (*yu shen* 雨神), as well as the thunder god (*lei shen* 雷神). Many of these spirits are regarded as heavenly governors of rivers and rains in both celestial and terrestrial realms.

Many early texts, including the *Classic of Mountains and Seas*, mention the worship of these popular deities by ancient people. In the *Classic of Mountains and Seas*, the River God, called Bingyi 冰夷, has a big face and rides two dragons. In his annotation, Guo Pu points out that the name Bingyi was actually Fengyi 馮夷. Citing the *Huainanzi* 淮南子, he notes that after Fengyi achieved "the Way," he hid himself in the great river,[46] that is, the Yellow River, and thus became the River God. Contemporary scholars have noted that the legend of Fengyi can be traced back to the *Zhuangzi*, where it is said that Fengyi traveled in the great river. Fengyi, whose name is Feng Yi, was originally from Hongnong 弘農 (modern Shangluo 商洛, Henan Province). He practiced alchemy and achieved transcendence, which allowed him to travel into the water of the Yellow River.[47] Wen Yiduo, in his philological approach to the names of the river god, proposed that the rainmaker's name Pingyi 屏翳 was a variant of Fengyi 馮翳, as it appears in the "Biography of Sima Xiangru 司马相如" in the *History of the Han Dynasty* (*Han shu* 漢書). It seems that the Chinese character *feng* 馮 is confused with the character *ping* 憑, which is pronounced the same as *ping* 屏. Fengyi 馮翳 refers to the River Uncle (Hebo Fengyi 河伯馮夷).

---

46 Yuan, *Shanhaijing jiaozhu*, 316.

47 Guo Qingfan 郭庆藩, *Zhuangzi jishi* 庄子集释 (Collective Annotations on Zhuangzi), annotated by Wang Xiaoyu 王孝雨 (Beijing: Zhonghua shuju, 1961), 249.

Fengyi 馮夷 is a homophone for Fengxi 封豨, meaning "hog," the water spirit of the Yellow River. It turns out that the rainmaker is a hog. Therefore, in the ancient legend, when a hog is seen wading into the waves, there will be rain.[48] This is the dynamic of the rainmaking process by a hog. Because of these similar but variant names in Chinese sources, Feng Yi and Fengxi, the hog came to be known as the River Uncle Fengyi, and the River Uncle who controlled the flood became the god of the river.[49]

Interestingly, Fengxi 封狶 is not only the water spirit of the Yellow River but is also identified as the Heavenly Hog in some sources, where it is written Fengshi 封豕. The "Astronomical Chapter" (*Tianguan shu* 天官書) in the *Records of the Grand Historian* (*Shiji*) clearly states that the lunar mansion called "Legs" (*kui* 奎) was also known as Tianshi 天豕 or Fengshi 封豕 (Heavenly Hog). The text says: "Legs is also called the Heavenly Hog; it dominates the rivers and streams" 奎曰封豕, 為溝瀆. In his commentary on this passage, Zhang Shoujie 張守節 (eighth century?) explains that "Legs is the heavenly treasury (*fu ku* 府庫), known as Tianshi, and also Fengshi, which rules over rivers and streams. It is a large constellation in the southwest, called *Tianshimu* 天豕目… When Mars comes close to it, there will be fears of flood for three successive years."[50] In the *Lüzhiyu* 履之豫 chapter of *Master Jiao's Forest of Changes* (*Jiaoshi Yilin* 焦氏易林), Jiao Yanshou 焦延壽 (first century BCE to first century AD?) links the phenomenon of pigs running into a marsh with the coming of a massive flood and identifies the "hog with white hooves" as the Heavenly Hog constellation.

Hence, in the Han, the Heavenly Hog constellation in the heavenly realm was regarded as the celestial office in charge of rivers and streams on the earth. According to this understanding, if the movement of the Heavenly Hog constellation went wrong, there would be a flood on the earth. The Heavenly Hog has become a celestial god that controls floods; it could also use floods to punish people on the earth.

Later on, the "Astronomical Treatise" (*Tianwen zhi* 天文志) in the *History of the Jin Dynasty* (*Jin shu* 晉書) continues this tradition of linking the Heavenly Hog in the celestial realm with the rivers and streams on the earth, while also adding more detailed information. The sixteen stars making up the

48 Wen, *Wen Yiduo quanji*, 48.

49 Lin Lisheng 林立勝 and Ye Chaolong 葉超龍, "Hebo shenhua youlai ji xingxiang chayi tantao 河伯神話由來及形象差異探討 (A Study on the Origin of the Duke of River and Its Various Images)," in *Shenhua yu wenxue lunwen xuanji 2012–2013* 神話與文學論文選輯 (*Selected Papers on Myths and Literature*) (Hong Kong: Department of Chinese Literature, Lingnan University, 2013), 82–89.

50 Sima Qian, *Shiji*, 1305.

Legs lunar mansion are said to be the "heavenly arsenal 天之武庫," known as Tianshi 天豕 and Fengshi 封豕. Their duties include threatening military force to avoid violence and ruling over the rivers and streams. A large constellation in the southwest was called Tianshimu 天豕目, or the "Grand General (Dajiang 大將)."[51] The "Astronomical Treatise 天文誌" in the History of the Sui Dynasty (*Sui shu* 隋書) also says: "If any of the stars in the Legs [lunar mansion] becomes brighter than usual, there will be a large quantity of rain."[52]

Another observance related to this theme is the connection between the Heavenly Hog and the movement of the moon. According to the "Astronomical Treatise" in the *History of the Sui Dynasty*, "When the moon is located to the north of the Hyades, there will be rain, and when to the south of the Hyades, no rain."[53] Later dynastic histories, such as the *History of the Song Dynasty* (*Song shi* 宋史), all repeat this statement, which has also been accepted in numerous divination manuals. Thus, through the association of the "Legs" lunar mansion with the Heavenly Hog constellation, and the role of controlling the rivers and streams, the pig was eventually transformed into the God of the Celestial River (*Tian he* 天河) who oversees rivers, streams, and floods, as well as water, everywhere.

So far, we have surveyed the textual sources, especially astronomical chapters in the official dynastic histories, to discuss the evolution of the pig into the Heavenly Hog, a celestial god. Popular proverbs also suggest that the observance of the celestial phenomena played a crucial role in linking the celestial phenomena and astronomical thought about the hog. As discussed in the previous section, the proverb about "the black pigs crossing the river" actually refers to the observance that clusters of black clouds drifting across the Milky Way is followed by rain. Although initially used allegorically in the folklores and literature for the agricultural uses, this phenomenon transforms the image of pig from a concrete terrestrial beast to an imaginary celestial creature.

We can thus see how the image of the "bathing pig" was first linked to rainfall and then took on supernatural qualities. As the Water God, the pig acquired the ability to summon the wind and the rain. Because

51 Fang Xuanling 房玄齡 et al., *Jin shu* 晉書 (History of the Jin Dynasty) (Beijing: Zhonghua shuju, 1974), 301.

52 Wei Zheng 魏徵, *Sui shu* 隋書 (History of the Sui Dynasty) (Beijing: Zhonghua shuju, 1973), 546.

53 Wei Zheng, *Sui shu*, 547; Yang Shiwen 杨世文 et al., *Yixue jicheng* 易學集成 (Integration of Yijing theory) (Chengdu: Sichuan daxue chubanshe, 1998), 1941.

the Heavenly Hog looks like the black pigs that exist in the celestial realm, it acquires the ability to control rivers and streams, as well as causing or controlling floods. It is also said to control thunder. A further development in the image of the pig is shown in a collection of Tang short stories entitled *Records of Strange and Mysterious Beings (Xuanguai lu* 玄怪錄), by Niu Sengru 牛僧孺 (779–848). It says that a General Wu 烏 (meaning "black") appeared as a black pig demon who can command the wind, rain, thunder, and hail. But Niu asserts the terrestrial origin of this general when he says, "General Wu, the god of the town, has been enshrined and worshiped by the local villagers for a long time. . . . Every year, a lady must be sent to him as a sacrifice to ensure that he will not cause trouble. If the lady is offered late, no matter how slightly, the local people are assailed by thunderstorms and hail."[54] In another collection, titled *Trivial Tales from North of Meng* (*Beimeng suoyan* 北夢瑣言), the ferocity of storms is depicted. The compiler Sun Guangxian 孫光憲 (900–968) relates that local officials in the region of Sichuan engaged in rituals, praying to the sow for rain in the fourth year of the Tianfu reign period (904). Their belief is described as follows: "When the songs and dances are at their height, black clouds suddenly rise from the pond, heaving in the cluster, accompanied by wild thunder and lightning. The clouds quickly turn dark, and the rain and hail immediately gush down."[55] This association of praying to the sow and the appearance of black clouds suggests that the black pigs formed into clouds and brought rain. The Heavenly Hog was seen as the god who controlled the thunder, lightning, and other natural phenomena.[56]

54 Niu Sengru 牛僧孺, *Xuanguai lu* 玄怪錄 (Records of Strange and Mysterious Beings) (Shanghai: Shanghai guji chubanshe, 1985), 24.

55 Sun, Guangxian 孫光憲. *Beimeng suoyan* 北夢瑣言 (*Trivial Tales from North of Meng*) (Shanghai: Shanghai gujichubanshe, 1981), 164.

56 The image of the Heavenly Hog as the Water God contributes to the archetypal image of Pigsy (Zhu Bajie 豬八戒) became so popular that it was incorporated into the famous novel *Journey to the West* (Xiyou ji 西遊記). The Heavenly Hog appeared as a military marshal who served in the heavenly court but was punished for falling on the ground as a troublesome black pig demon; see Yao Lijiang 姚立江 and Pan Lanxiang 潘蘭香, *Renwen dongwu: Dongwu fuhao yu zhongguo wenhua* 人文動物: 動物符號與中國文化 (Humanistic Animals: The Animal Signs and Chinese Culture) (Haerbin: Heilongjiang renmin chubanshe, 2002), 357; Suzuki Yūiki 鈴木陽一, "Seiūki niokeru jinbutsu keishū no seikentō 2 Chūhakka to Songoku 西遊記における人物形象の再檢討 (2) 豬八戒と孫悟空 (A Reexamination of the Images of Figures in the Journey to the West 2: Zhu Bajie and Sun Wukong)." *Gengo bunka kenkyū* 言語文化研究 (*Studies on Languages and Cultures*) 1, no. 1 (1981): 123–50.

Conclusions

In sum, observing cyclical and seasonal natural phenomena, especially climate change and the behavior of animals, has a long history throughout the world. This chapter argues that the tendency for pigs to take to the wet environment especially the water led the ancient people in China who observed it to develop a divination technique of predicting rainfall and even flood. This technique combined with the concept of the pig deity controlling rainfall and flood and developed the cult of the pig. China's unique continental monsoon climate, which is characterized by hot, wet summers seems to provide the reason for the pig's tendency to take to the water in order to cool down its body, which sharply contrasted to the pigs' wallowing in mud in other regions of Asia. Although the particular details of the climate cannot help to explain why the cult of the pig did not develop in the West, it does help to explain why it became established in ancient China. Ancient people used their observations of the behavior of pigs to predict the weather in China, and the connection between this behavior and the weather was seen as a mysterious or magical phenomenon. Such phenomena were easily able to become the foundations for religious beliefs, gradually leading to the worship of pigs as deities. They were even sometimes used in religions and folk beliefs as evidence to prove the existence of spiritual beings or forces.

The rise of the cult of the pig in ancient China had a particular spatial and temporal context. The climate and natural environment defined by this context also contributed to this rise. The bathing pig and the black pig crossing the river were originally phenological or climatological phenomena. However, with China's unique climate—its wet summers and dry winters, as the rainfall lines show—these phenomena were sometimes successful in predicting rainfall. When they were successful, the effectiveness of the method of observing the behavior of pigs to predict the weather was confirmed, at least for premodern Chinese observers. Ancient China's divination techniques not only involved recording the existence of certain phenological phenomena but also incorporated some ancient philosophical concepts such as the *Yin-Yang* and Five-phase theories. Thus, they are distinctly different from modern phenology.

A further implication is what Chinese ecocriticism we can learn about from this case study. This study reveals a big difference between ancient Chinese ecological issue and modern wicked problem based on science and technology.[57]

57 Craig Thomas and David Pijawka, "Studying American Naturalists for Transdisciplinary Solutions to Today's Wicked Problems," *ISLE: Interdisciplinary Studies in Literature and Environment* 27, no. 2 (2020): 221–42.

Traditional agriculture is often considered as a contract between humans and nature.[58] Yet modern biochemical and genetic technology have almost entirely taken over meat industry in most developed and many developing countries. Rearing pigs is no longer closely connected with natural environment and climate. Pigs and some other animals are more or less regarded as the raw materials for producing meat. They are no longer aware of nature's rhythms. Naturalists would not have an opportunity of observing the behaviors of pigs in their responses to the changes of environment and climate. In contrast, as naturally trained "naturalists," ancient Chinese people observed pigs from a holistic perspective and acknowledged the unique position of pigs in their material and spiritual life. The pig was not regarded as a main companion in the terrestrial realm for humans but also was viewed as a medium between celestial power and natural phenomenon on the earth. Shall today we rebuild these connections between animal and environment, between animal and humans, and between heaven and earth? This chapter might have offered some inspirations for our readers.

58 Roger Straughan, *Ethics, Morality and Animal Biotechnology* (Swindon: Biotechnology and Biological Sciences Research Council, 1999).

Chapter 4

ZOOMANCY IN MEDIEVAL CHINA

Introduction

In the historical context of ancient China, there was no single text solely devoted to zoomancy, if we understand zoomancy, or divination by animals, as a method of prediction by observing animal behaviors.[1] As a prevalent divination technique in premodern China, zoomancy was popular among both elite and lower classes. However, it was not the primary technique utilized by the royal diviners (*taibu* 太卜) who served in the court to assist the rulers in making decisions on political and military affairs. It was more common in random events when the elite and lower classes observed phenomena and then associated those phenomena with unusual animal behaviors. The authority of interpreting these unusual animal behaviors did not necessarily reside with professional diviners who held appointments in the court, but rather with any civil and military officials who might know about zoomancy. In other words, unlike oracle bone divination, zoomancy was not professionalized or monopolized by professional diviners in ancient China.

The philosophical and intellectual foundation of zoomancy was correlative thinking, which refers to the ancient belief that natural phenomena were associated with human behaviors and social events. For example, rulers believed that their statecraft and governance were connected with the movements of the cosmos, including various natural phenomena and animal behaviors. If their governance was virtuous, there would be auspicious signs from natural phenomena, and animals would respond with certain behaviors as a verification. If a ruler was about to die or a regime was about to fall, there were bad omens from nature, including unusual animal behaviors.

1 Predictions associated with dead animals or parts thereof, such as bones, is not zoomancy; this is covered in Kenichi Takashima's chapter on bone divination or osteomancy in the *Handbook of Chinese Divination* (ed. by Stephen Kory). Prognostication associated with the pronunciation of animal names, such as how the word parrot predicted the fates of Gao Huan 高歡 and Wu Zetian 武則天, will not be discussed in this chapter either. I have discussed it in *In the Land of Tigers and Snakes: Living with Animals in Medieval Chinese Religions*, chapter 6.

Zoomancy in China has a long history, which can be traced back to the Spring and Autumn period (771–476 BCE), at the latest. According to Zuo Qiuming's 左丘明 (556–451 BCE) alleged fourth century BCE *Zuo's Commentary on the Spring and Autumn Annals* (*Chunqiu Zuozhuan* 春秋左傳), in 555 BCE, during a war between the states of Jin 晉 and Qi 齊, Shi Kuang 師曠 (ca. 572–532), a talented musician who served the Marquis of Jin 晉侯, heard the joyful sound of birds and told the Marquis that this sort of sound indicated the withdrawal of the Qi troops. Indeed, the Qi troops fled that night.[2] Another work attributed to Zuo, the *Conversations of States* (*Guoyu* 國語) also states that *yuezhuo* 鸑鷟, a mythic phoenix-like bird with red eyes, sang on Mount Qi 岐山 (western Shaanxi 陝西), indicating the rise of the Zhou 周 state.[3] Both cases refer to political and military phenomena. The interpreters of such zoomancy [zoomantic signs] were officials who served in the court, which might also be why both cases appeared in written documents.

Records concerning zoomancy flourished during the Han dynasty (202 BCE–220 CE). Both official histories such as the *Book of the Han Dynasty* (*Hanshu* 漢書) and numerous apocryphal texts (*weishu* 緯書) on prophecy and apocrypha (*chenwei* 讖緯), as well as Confucian writings such as the *Family Sayings of Confucius* (*Kongzi jiayu* 孔子家語), recorded many instances about divination by animals. Such widespread zoomancy in the sources indicates that the zoomancy practiced in the Han period was not limited to particular social classes. In the *Book of the Han Dynasty*, a section entitled the "Treatise on the Five Phases" (Wuxing zhi 五行志) detailed various natural and cosmic phenomena and their relationships to social and political events.[4] The apocryphal texts recorded even more phenomena for prognostication. These two genres laid the textual foundation for later generations to continue writing about zoomancy in imperial China. Since the Han dynasty, many official histories from the *Book of the Jin Dynasty* (*Jinshu* 晉書) to the *History of the Ming Dynasty* (*Mingshi* 明史) also included a treatise on the five phases for recording auspicious and unfortunate omens predicting the rise and fall of various regimes.[5] As these texts reveal in the later periods, correlative cosmology functioned in various divination techniques throughout the ages.[6]

2 *Chunqiu Zuozhuan zhu* 3.1038 (襄公十八年).

3 *Guoyu* 12.25 (周語上).

4 *Hanshu* 27A.1315–27C.1522 (五行志).

5 E.g., *Jinshu* 27.799–29.910 (五行志); *Mingshi* 28.425–30.512 (五行志).

6 For a general survey on the connected tradition of divination and correlative cosmology in ancient China, see Richard J. Smith, *Fortune-Tellers and Philosophers: Divination in Traditional Chinese Society* (Boulder: Westview Press, 1991), 1–47.

Other treatises on the five phases were compiled by private scholars, such as *Master Qiao's Treatise on Five Phases* (*Qiaozi wuxing zhi* 譙子五行志) compiled by Pu Yangxia 濮陽夏 during the Tang dynasty (618–907), which also recorded numerous zoomantic practices.[7]

New zoomancy trends appeared in the early medieval periods, especially during the Northern and Southern Dynasties (386–581). One was associated with the rise of new literary writings, especially "strange writings" (*zhiguai* 志怪), which recorded many stories about abnormal natural and social phenomena, including the behaviors of animals and their relationships to social events. Another trend was the spread of Buddhism, which integrated the Buddhist theory of karmic retribution into the prognostication practices of local communities.[8] In Buddhism, both animals and humans were regarded as sentient beings who would experience birth, death, and rebirth within six realms. Therefore, their behaviors might contribute similarly to the prognostications. The third trend was that the thought and practice of zoomancy in medieval China sometimes depicted entangled relations between the Han ethnic group and nomadic groups, and between Buddhism and Daoism. For example, according to the *Book of the [Liu-]Song Dynasty* (*Songshu* 宋書), when Liu Yu 劉裕 (363–422) was attacking Guanggu 廣固 (central Shandong) in 399, a giant goose-like bird flew into his military tent. One of his officials, Hu Fan 胡藩 (372–433), interpreted this abnormal phenomenon as an auspicious omen that the vicious enemy would be defeated because the blackbird symbolized the barbarians.[9] These trends continued through to the Tang and the Song (960–1279) dynasties as recorded in numerous "tales of the strange" (*chuanqi* 傳奇) and fortune-telling handbooks (*xiangshu* 相書). After the Tang dynasty, more texts on animals and the agricultural uses of animals appeared. Some divinations by animals were recorded in the "random notes" (*biji* 筆記) of the Song dynasty. In late imperial China, zoomancy even appeared in popular proverbs and sayings to help farmers predict weather changes pursuant of optimizing agricultural production.

Zoomancy traditions in ancient China represent an underdeveloped field in contemporary scholarship. Some general surveys on the history of divination in ancient China neglect zoomancy. Monographs, collected volumes, and articles

7 *Qiaozi wuxing zhi* 3: 585–751.

8 Zoomancy is briefly mentioned in Yang Gang and Christoph Anderl, "Prognostication in Chinese Buddhist Historical Texts: The *Gaoseng zhuan* and the *Xu Gaoseng zhuan*," *Acta Orientalia Academiae Scientiarum Hungaricae* 73.1 (2020): 1–45.

9 *Songshu* 50.1444 (胡藩傳).

might briefly mention trivial cases on divinations by animals but remain focused on more visible divination skills, such as bone divination. This scholarly interest seems to correspond to the marginal position of zoomancy in ancient China. After all, zoomancy was not a significant divination technique practiced by Chinese officials or court diviners. Specifically, there is no single monograph in any Western language dedicated to zoomancy in premodern China, though some publications mention zoomancy. For instance, Martin Kern introduces observations of natural phenomena, including animal behavior in the *Master Lü's Spring and Autumn Annals* (*Lüshi Chunqiu* 吕氏春秋). Still, this has nothing to do with zoomancy.[10] In her comparative studies on divination and predication in early China and Greece, Lisa Raphals notes that some texts mention a *Physiognomizing the Six Kinds of Domestic Animals* (*Xiang liuxu* 相六畜), but in ancient physiognomy, animal comparisons were fanciful and rhetorical.[11] In his monumental work on divination in the Han dynasty, Michael Loewe notes the common practice of divination by shells, bones, and stalks but not animals, though he pays attention to the hybrid images of humans and animals in ancient Chinese art and literature as well as the cult of the dragon.[12] Some scholars have focused on manuscripts from Dunhuang, but several articles in a volume of collected papers edited by Marc Kalinowski only briefly mention animals as auspicious signs on some occasions in some manuscripts.[13] Scholars in East Asia have published numerous books and articles on divination. Some of the sections in these publications deal with zoomancy,[14] but there is no comprehensive monograph on the subject.

---

10 Martin Kern, "Early Chinese Divination and Its Rhetoric," in *Coping with the Future: Theories and Practices of Divination in East Asia*, ed. Michael Lackner (Leiden: Brill, 2017), 268.

11 Lisa Raphals, *Divination and Prediction in Early China and Ancient Greece* (Cambridge: Cambridge University Press, 2013), 142–44.

12 Michael Loewe, *Divination, Mythology and Monarchy in Han China* (Cambridge: Cambridge University Press, 1994).

13 Marc Kalinowski, ed., *Divination et société dans la Chine médiévale: Étude des manuscrits de Dunhuang de la Bibliothèque nationale de France et de la British* (Paris: Bibliothèque nationale de France, 2003).

14 Huang Zhengjian 黃正建. *Dunhuang zhanbu wenshu yu Tang Wudai zhanbu yanjiu* 敦煌占卜文書與唐五代占卜研究 (A study on the divination manuscripts from Dunhuang and the divination culture in the Tang and Five Dynasties). Beijing: Zhongguo shehui kexue chubanshe, 2014; Yan, Yaozhong 嚴耀中. "Lun zhanbu yu Sui Tang fojiao de jiehe 論占卜與隋唐佛教的結合 (On the Combination between Divination and Buddhism in the Sui and Tang Dynasties)." *Shijie zongjiao yanjiu* 世界宗教研究 4 (2002): 30–37; Kōno Kimiko 河野喜美子. *Nihon ryōiki to Chūgoku no denshō* 日本靈異記と中国の伝承 (*The Record of Miraculous Stories in Japan and the Chinese Transmissions*) (Tōkyō: Benseisha, 1996).

Historical Sources for Understanding Zoomancy

The primary sources of zoomancy in ancient China are official histories, apocryphal texts, strange writings, tales of wonder, and encyclopedias of divination. Two groups of sources distinguish themselves from numerous texts that only briefly mention zoomantic practices or random cases. These two groups include treatises on the five phases in the official histories and encyclopedias of divination. Before the popularity of official histories, earlier histories such as *Zuo's Commentary on the Spring and Autumn Annals* and *Conversations of States* mentioned zoomantic practices several times, mostly on how strange animal behaviors indicated political events and military campaigns. Ban Gu's (32–92 CE) "Treatise on the Five Phases" in the *Book of the Han Dynasty* officially started the tradition of recording cases on zoomancy in the political and civil domains of different states. From the *Book of the Han Dynasty* to the *History of the Ming Dynasty*, numerous official histories record zoomantic practices. These practices are not grouped together or categorized explicitly. Because many divinations and prophecies are listed chronologically, zoomancy and other prophecies tend to be mixed together.[15] Some cases appearing in the treatises on the five phases were selected from *Zuo's Commentary on the Spring and Autumn Annals* or from lost sources. Prognostications on disasters and anomalies in recently excavated texts written on bamboo strips belonging to the old version of the *Book of Changes* (*Yijing* 易經) exhibit close similarities in content to the treatises on the five phases in the early official histories.[16]

Many apocryphal Han dynasty texts focused on divinations and prophecies.[17] Some of these apocryphal texts are catalogued in the "Treatise on

---

15 Contemporary scholarship does not focus on zoomancy, but rather on the relations between cosmology and political culture. For studies of sections on the five phases in the Chinese dynastic histories, see Wang, *Cosmology and Political Culture in Early China*; Michael Nylan, "The 'Treatise on the Wuxing' (Wuxingzhi 五行誌)," in *Monographs in Tang Official Historiography: Perspectives from the Technical Treatises of the History of Sui (Sui shu* 隋書), eds. Daniel Patrick Morgan, Damien Chaussende, and Karine Chamla (Dordrecht: Springer Nature, 2019), 181–236.

16 Edward L. Shaughnessy, *Unearthing the Changes: Recently Discovered Manuscripts of the Yi Jing (I Ching) and Related Texts* (New York: Columbia University Press, 2014), 142.

17 For examples, see Jack L. Dull, "A Historical Introduction to the Apocryphal (Ch'an-Wei) Texts of the Han Dynasty." PhD dissertation, University of Washington, 1966; Hans van Ess, "The Apocryphal Texts of the Han Dynasty and the Old Text/New Text Controversy," *T'oung Pao* 85, nos. 1–3 (1999): 29–64; Tiziana Lippiello, *Auspicious Omens and Miracles in Ancient China: Han, Three Kingdoms and Six Dynasties* (Sankt August: Monumenta Serica Institut, 2001); Lü Zongli, *Power of the Words: Chen Prophecy in Chinese Politics, AD 265–618* (Bern and New York: Peter Lang, 2003); Zhao Lu, *In Pursuit of the Great Peace: Han Dynasty Classicism and the Making of Early Medieval Literati Culture* (Albany: State University of New York Press, 2019).

Arts and Literature" (Yiwen zhi 藝文志) of the *Book of the Han Dynasty*, yet none of them are fully extant. As we have already seen from the above discussions, several important apocryphal texts such as Jing Fang's 京房 (ca. 76–37 BCE) commentary on the *Book of Changes*, the *Spring and Autumn Explanatory Kong Chart* (*Chunqiu yan Kong tu* 春秋演孔圖), and the *Extended Mystical Tally of the Classic of Filiality* (*Xiaojing yuanshen qi* 孝經援神契) did not survive in their entirety. Still, fragments are preserved in Tang encyclopedias of divination like *Treatise on the Auspicious Signs of Heaven and Earth* (*Tiandi ruixiang zhi* 天地瑞祥志) and the *Classic of Divination Compiled in the Kaiyuan Period* (*Kaiyuan zhanjing* 開元占經).

These apocryphal texts from the Han preserved numerous divination stories. Zoomancy only represented a small fraction of the colorful divination culture manifested in these apocryphal texts. Jing Fang was the Prefect of Weijun 魏郡 during the Han dynasty. The first two fascicles of his three-fascicle commentary on the *Book of Changes* interpret the text's combinations of trigrams (*gua* 卦), while the third introduces divination methods in terms of the relations among heaven, earth, humans, and demons, focusing on the interpretation of abnormal events.[18] Although the original version of this commentary was lost, it has been widely cited in many historical works. For example, it is cited more than seventy times in the *Book of the Han Dynasty* "Treatise on the Five Phases,"[19] and references in the Tang encyclopedia *Treatise on the Auspicious Signs of Heaven and Earth* relate various zoomantic practices about abnormal events. References to the *Spring and Autumn Explanatory Kong Chart* collected in the *Treatise on the Auspicious Signs of Heaven and Earth* and the *Classic of Divination Compiled in the Kaiyuan Period* also mention numerous zoomantic issues. Given the various passages cited in these divination works, a detailed philological study might reveal the transformations of particular zoomantic issues.

The most comprehensive collections on zoomantic practices are fascicles 18 and 19 of the *Treatise on the Auspicious Signs of Heaven and Earth* and fascicles 115–120 of the *Classic of Divination Compiled in the Kaiyuan Period.* Both texts appeared in the Tang, and both were compiled based on the same original sources. Their contents seem to overlap but differ slightly due to variant characters. The former was discovered in Japan rather than in China. Numerous scholars have discussed the authorship of the *Treatise on the Auspicious*

18 Rafe de Crespigny, *Portents of Protest in the Later Han Dynasty: The Memorials of Hsiang K'ai to Emperor Huan* (Canberra: Australia National University Press, 1976), 70–71.

19 Michael Loewe, *Dong Zhongshu, a "Confucian" Heritage and the Chunqiu Fanlu* (Leiden: Brill, 2011), 125.

Signs of Heaven and Earth and its actual place of compilation. It was unknown in any Chinese bibliographical record until manuscripts dated to 1686 were found in the Sonkeikaku Library 尊經閣文庫 (Tokyo). These manuscripts only include fascicles 1, 7, 12, 14, 16, 17, 18, 19, and 20; claiming that Sa Shouzhen 薩守鎮 (fl. 666 CE) completed the work in the third year of the Tang Linde 麟德 reign period (666 CE).[20] Fascicle 18 collects zoomantic stories on birds, with a much shorter corresponding version appearing in fascicle 115 of the *Classic of Divination Compiled in the Kaiyuan Period.* Fascicle 19 corresponds to fascicles 116–120 of the *Classic of Divination Compiled of the Kaiyuan Period*, but there are differences in terms of the order of many sections. The *Treatise on the Auspicious Signs of Heaven and Earth* cites numerous short and fragmentary passages from various Han apocryphal texts and tales about anomalies and marvels from the early medieval period. The citations are not always accurate. They should be read against the corresponding sections in the *Classic of Divination Compiled in the Kaiyuan Period* and their cited sources whenever they are available.

Another important collection is *Master Qiao's Treatise on Five Phases.* There is no specific date for the production of this five-fascicle collection, but it appeared first in the "Treatise of Arts and Literature" in the *New History of the Tang Dynasty* (*Xin Tangshu* 新唐書).[21] One Ming (1368–1644) and one Qing (1644–1912) dynasty copy are preserved at the National Library in Beijing. In each of its five fascicles, two or three sections are devoted to zoomancy. For instance, fascicle one has sections on dragons, snakes, and fish. Fascicle two has sections on birds, vultures, pheasants, and roosters. Fascicle four has sections on oxen, horses, dogs, hounds, and mice. The last fascicle has sections on turtles. Many of the records on animal divination in this collection can be also found in Sa Shouzhen's collection, but they tend to be more concise.

Besides traditional sources, manuscripts from Dunhuang also offer some information about medieval Chinese zoomantic practices. Some manuscripts are divination or ritual handbooks, including books for interpreting dreams, daybooks, and diagrams. Not all sources mention zoomancy. While one of

20 Another copy of this text, dated 1932, is preserved at the Institute of Humanities Research at Kyoto University. It was reprinted in Gao Keli ed., *Xijian Tangdai tianwen shiliao sanzhong.* On the textual history of this piece, see Mizuguchi Motoki 水口幹記, *Nihon Kodai Kaiseki juyō no shiteki kenkyū* 日本古代漢籍受容の史的研究 (A study on the history of the acceptance of Classical Chinese texts in ancient Japan) (Tokyo: Kyūko Shoin, 2005), 177–208. Other studies on this text can be found in Nawa, ed., *Higashiajia shisō Bunka no kisō kōzō.* Numerous articles in East Asia have offered detailed annotations and studies on this text, which I cannot exhaustively cite here.

21 *Xin Tangshu* 59.1558 (藝文志).

the most interesting manuscripts from Dunhuang is the *Diagrams of Spectral Prodigies of White Marsh* (*Baize jingguai tu* 白澤精怪圖), it does not represent what we understand as zoomancy, but features animal-like demonic marvels.[22]

The Categorization of Zoomancy

Bibliographically speaking, numerous zoomantic practices appear in two sets of textual documents in the premodern Chinese bibliographical system, namely official documents and popular documents. Official documents are often compiled by government officials and served political governance objectives. These documents include official histories, annals, and institutional documents, as well as Confucian classics. The authorship or editorship of these documents is often explicit. Popular documents include strange writings, tales, legends, and religious documents from Buddhism and Daoism, which are seldom compiled under the sponsorship and support of the court or government. They might be compiled by literati, but the original authorship of most of them is unknown. They do not serve political motivations and objectives, reflecting how people understand and interpret correlative cosmology by observing animal behaviors at the local or non-bureaucratic level. However, in the Song period (960–1279), governments also compiled large-scale encyclopedias, such as the *Finest Blossoms in the Garden of Literature* (*Wenyuan yinghua* 文苑英華), *Extensive Records of the Taiping Xinguo Period* (*Taiping guangji* 太平廣記), *Imperial Overview from the Taiping Xingguo Period* (*Taiping yulan* 太平御覽), and *Primary Turtle of the Records Office* (*Cefu yuangui* 冊府元龜). Thus, both official and popular documents are collected and recategorized to fit the needs of the rulers.

More specifically, though many cases about divination by animals appear in the biographies of officials, zoomantic practices are more commonly described in the treatises on the five phases in official histories. It should be noted that, even within these treatises, there is no specific section which focused exclusively on zoomantic phenomena. Instead, some cases are recorded because the compilers view them as extraordinary events within the traditional framework of correlative cosmology.

22 For details, see Kalinowski, ed., *Dunhuang de la Bibliothèque Nationale de France et de la British*, 455–58; Huang, *Dunhuang zhanbu wenshu yu Tang Wudai zhanbu yanjiu*; Donald Harper, "The Textual Form of Knowledge: Occult Miscellanies in Ancient and Medieval Chinese Manuscripts, 4th Century BCE to 10th Century CE," in *Looking at It from Asia: The Processes that Shaped the Sources of History of Science*, ed. Florence Bretelle-Establet (Dordrecht: Springer, 2010), 37–80. There are also numerous articles written in Chinese and Japanese.

The recording and compiling of these cases seem to be connected to the bureaucratic system in ancient China. According to the early fourteenth century *Comprehensive Investigations Based on Literary and Documentary Sources* (*Wenxian tongkao* 文獻通考), the establishment of an official position in charge of divination can be traced back as early as the Shang dynasty (ca. 16001046), wherein this role was given the title of the "grand diviner" (*taibu* 太卜), one of six critical positions in the court. This position continued through the Zhou (ca. 1046–256 BCE), but was renamed "director of [the office of] the grand diviner" (*taibu ling* 太卜令) during the Qin (221–206 BCE) and Han. In the Later Han dynasty (25–220 CE), this position was incorporated into the office of the "grand historian" (*taishi* 太史). During the Northern Wei (386–535), the person in charge of divination was known as the "erudite of divination" (*bu boshi* 卜博士), but various titles were used to designate this position during the Northern Dynasties (386–581). In the Sui (581–618) and Tang dynasties, a new office was established, and a director and deputy were appointed. Then in the Song dynasty, this office was incorporated into the Astronomical Bureau (*Sitian jian* 司天監).[23] Supposedly, the remit of the grand diviner spanned all divinations.

As I indicated earlier, zoomantic practices seem not to be a major divination technique in ancient Chinese courts. Therefore, the grand diviner is seldom the chief officer who appears in the numerous zoomantic cases in institutional and popular documents. Various officials and literati who happen to have experience observing unusual animal behaviors, however, might stand up to interpret these phenomena for the rulers or local communities. For example, according to an apocryphal Han dynasty text entitled *Matched Prophecy and Realized Divination of the Classic of Documents* (*Shangshu zhonghou* 尚書中候), Duke Huan of Qi 齊桓公 (d. 643) once plans to offer sacrifices to Mount Tai 太山 (west-central Shandong), but Guan Zhong 管仲 (720–645 BCE) tells him that the sacrifice should not be offered unless a flatfish (*bimuyu* 比目魚), a phoenix, and a kirin (*qilin* 麒麟) appear.[24] In this case, Guan Zhong is not a professional grand diviner but a knowledgeable chancellor who served Duke Huan and had knowledge of various natural phenomena and correlative social changes. In other words, the grand diviner did not have a monopoly on zoomantic knowledge in ancient times; it seems that many political and military officers were familiar with many zoomantic issues.

Zoomantic knowledge from the medieval period can be found in encyclopedic collections of divinations and prophecies that recorded

23 *Wenxian tongkao* 55 (太卜署).

24 *Shangshu zhonghou* 13.

auspicious and unfortunate omens for the political events and rulers, such as the 666 CE *Treatise on the Auspicious Signs of Heaven and Earth* and the 729 *Classic of Divination Compiled in the Kaiyuan Period.* In these encyclopedias, zoomantic cases were broadly categorized into two groups, namely "bird divination" (*qinzhan* 禽占) and "beast divination" (*shouzhan* 獸占). The compilers often did not have specific scientific or ethnobiological knowledge for classifying animals. For example, fascicle 18 of *Treatise on the Auspicious Signs of Heaven and Earth* is mostly devoted to birds, including both mythic birds like the phoenix and *luan* 鸞 bird, and real nonhuman birds in nature such as eagles, vultures, sea birds, geese, and pheasants. But this fascicle also concludes with some insects, such as butterflies, cicadas, flies, bees, locusts, and ants; and surprisingly, fish, turtles, crabs, and toads. Fascicle 19 collects stories about legendary beasts, such as the kirin, the dragon, and mythic horses; as well as both domestic and wild beasts like elephants, horses, buffaloes, sheep, dogs, tigers, wolves, bears, pigs, deer, foxes, rabbits, monkeys, rhinoceroses, mice, and snakes.[25] Therefore, it only reflected a rough knowledge and imagination of the animal kingdom in the medieval period, which was different from the traditional classification of animals in the ancient Chinese lexicon *Erya* 爾雅 or *Approaching Elegance* and in medieval Chinese Buddhist classification systems. The former classified animals into four types, namely birds, beasts, insects, and fish, whilst the Buddhist masters classified animals into three groups: domestic animals, wild animals, and animals prohibited by the monastic code.[26]

Methods of Zoomancy in Medieval China

In ancient China, zoomantic practices included both imaginary mythical animals and real nonhuman animals. Textual sources record both types of animals. Even for real nonhuman animals in nature, there is no scientific identification of a particular species. Animals were most often mentioned by

---

25 See, respectively, *Tiandi ruixiang zhi* 18 and 19. It should be noted that different animals might be active in different regions across China. For instance, horses, sheep, and cattle were the most important domestic animals in north China in the medieval period; see Keith Knapp, "The Use and Understanding of Domestic Animals in Medieval Northern China." *Early Medieval China* 25 (2019): 85–99; Huaiyu Chen, *In the Land of Tigers and Snakes*, chapter 1.

26 For the former, see Roel Sterckx, "Animal Classification in Ancient China," *East Asian Science, Technology and Medicine* 23 (2005), 96–123; for the latter, see Chen Huaiyu, "A Buddhist Classification of Plants and Animals in Early Tang China," *Journal of Asian History* 43, no. 1 (2009): 31–51.

their generic names. For example, in many ancient Chinese sources, *niu* 牛 (oxen) might refer to either buffalo or cow, which cannot be determined in most cases even where there is a historical context. In the following discussions, based on the most comprehensive collection of zoomantic practices in the *Tiandi ruixiang zhi* or *Treatise on the Auspicious Signs of Heaven and Earth*, I will categorize animals into the following basic types: mythical birds, other birds, insects, mythical beasts, domestic beasts, amphibians, and fish.

In ancient China, observers regard animal behaviors as auspicious and unfortunate omens of political and social changes. These animal behaviors are connected to observers' understandings of spatiality, which means that the interpretation of these behaviors depends entirely on where they appear in the human landscape. The spatial concept in ancient Chinese zoomancy is manifested as a bifurcation: the court (*chao* 朝) or government space, and the field (*ye* 野) or rural area. In other words, the court or the government space is where political power, authority, legitimation, and the ruling class resides. The field is a politically marginal and powerless space that the court rules. How animals behave in the court and the field would result in different interpretations, depending on how they effected the legitimacy of the court and ruling class.

In the classification system of the *Erya*, the bird refers to a flying animal with two feet and feathers.[27] According to Han scholar Jing Fang's commentary on the *Book of Changes*,[28] extraordinary birds represent bad omens. For instance, if a bird has the patterns of a beast or the shape of one of the six domestic animals, it could predict the breakout of war.[29] Many apocryphal texts discussed the behaviors of mythical birds, such as the phoenix and *luan* 鸞 bird. These mythical birds are created in ancient art and literature. They vary in terms of their shape and the patterns of their feathers. Since the phoenix is usually connected with royal power and authority, its arrival in the court is regarded as an auspicious sign of great peace resulting from virtuous governance. The *luan* bird is a similarly auspicious bird.[30] In the Han apocryphal text *Charts of the Spring and Autumn*'s *Explanations of Confucius*

27 *Erya* 17. 340 (釋鳥).

28 On Jing Fang, see de Crespigny, *Emperor Huan and Emperor Ling*, 380.

29 *Tiandi ruixiang zhi* 352. Since the original version of Jing Fang's text *Jingshi yizhuan* 京氏易傳 is lost, in this paper I only cite its passages preserved in *Tiandi ruixiang zhi*.

30 For the *luan* bird, see James M. Hargett, "Playing Second Fiddle: The Luan-Bird in Early and Medieval Chinese Literature," *T'oung Pao* 75, nos. 4 & 5 (1989): 235–62.

(*Chunqiu yan Kong tu* 春秋演孔圖),[31] it states that a *luan* bird would appear in the field if the Son of the Heaven selected and promoted officials because of their virtuous governance. In his *Encomia on the Illustrations in the Classic of Mountains and Seas* (*Shanhai jing tuzan* 山海經圖讚), Guo Pu 郭璞 (276–324) notes that the arrival of the *luan* bird foretells the safety and peace of society. Besides the phoenix and *luan* bird, the behaviors of other divine birds (e.g., arriving, flying, singing, dancing) could deliver signs about the rulers and their governance.[32] The *Classic of Mountains and Seas* mentions a mythic bird with a human face and giant tail, whose appearance and singing predict tremendous drought. The appearance of another mythical bird, with one foot and a pigtail, is connected with a severe pandemic across the country.[33] Some other birds might also indicate the rise of war (often invasion by nomads), plagues, floods, or droughts that would damage or even destroy the country.

Real birds also serve as auspicious and bad omens. Storks, for example, are not domesticated and should be found in the field. If they were found to assemble in the court, it would indicate terrible consequences. According to the *History of the Jin Dynastic Revival* (*Jin zhongxing shu* 晉中興書), in 391 CE, a group of storks appeared in the Taiji Hall 太極殿 of the royal palace, and two years later, storks assembled in the Eastern Palace 東宮, where the crown prince was residing. These abnormal behaviors of storks indicate the miserable fate of the crown prince at that time, Sima Dezong 司馬德宗 (382–419; r. 397–418), who was later killed by Liu Yu 劉裕 (363–422; r. 420–422), marking the end of the Eastern Jin dynasty (317–420).[34] The mid-fourth-century *Book of the Jin Dynasty* states that a sinkhole formed

31 This text is cited as *Chunqiu Kong yan tu* 春秋孔演圖 (Chart of Confucius's expositions in the Spring and autumn [annals]) in *Tiandi ruixiang zhi* 18. Zhao Lu translates the title as *Annals of Diagrams Elaborated by Confucius* in his *In Pursuit of the Great Peace*, 238 note 12. The same work appears as both *Chunqiu Kong yan tu* and *Chunqiu yan Kong tu* in *Kaiyuan zhanjing*. It also appears as the *Chunqiu wei yan Kong tu* 春秋緯演孔圖 (Chart of the Spring and autumn apocrypha expounding on Confucius) in the Qing dynasty *Yuhan shanfang ji yishu* 玉函山房輯佚書 (Collection of lost texts from the Jade Bookcase Mountain Studio). For more on the *Chunqiu wei yan Kong tu*, see Lionel M. Jensen, "The Genesis of Kongzi in Ancient Narrative: The Figurative as Historical," in *On Sacred Grounds: Culture, Society, Politics, and the Formation of the Cult of Confucius*, ed. Thomas A. Wilson (Cambridge, MA: Harvard University Press, 2002), 206; Paul R. Goldin, "Legends of Confucius," in *Hawaii Reader in Traditional Chinese Culture*, eds. Victor H. Mair, Nancy S. Steinhardt, and Paul R. Goldin (Honolulu: University of Hawai'i Press, 2005), 234.

32 *Shanhai jing tuzan*, 7 (鸞鳥).

33 *Shanhai jing* 5.197 (中次十經).

34 *Jin zhongxing shu* cited in *Tiandi ruixiangzhi* 18.

in an area within the northeastern corner of Luoyang 洛陽 and two geese appeared in it. One was black and could fly, but the other was white and could not fly. The emperor consulted with erudite individuals but could not find an interpretation for this event. Dong Yang 董養, a locally recommended official, told the emperor that the black goose referred to nomadic barbarians who would come to attack and conquer Luoyang, and the white goose which could not fly indicated a state taboo.[35]

Unusual behaviors of domestic birds such as roosters and hens could also serve as auspicious and bad omens. For example, according to Jing Fang's commentary on the *Book of Changes*, when domestic roosters fought with wild birds, it would indicate the country would fall into chaos. Subjects would kill their lords and civil war might break out. If domestic roosters resided together with wild pheasants, it would indicate that a despicable person would become a lord. If a hen did not lay eggs or gave birth to babies with shapes of six domestic beasts, it would indicate a war disaster. If the babies looked like fish, it would mean a flood disaster would occur.[36] Many stories also tell that disaster would occur if a hen transformed her gender and became a rooster. For instance, in the *Book of the Later Han Dynasty* (*Hou Hanshu* 後漢書) "Treatise on the Five Phases," when Emperor Ling 靈帝 (156–189; r. 168–189) ruled the country, an unusual phenomenon about a hen's transformation into a rooster was reported. Later on, the Yellow Turban (*huangjin* 黄巾) rebellion brought about the collapse of the Han empire.[37]

Although many birds serve as signs for predictions and verifications of political and military changes, some birds serve as signs to indicate family issues. Crows were usually associated with black, but an appearance of white crows might indicate auspiciousness. According to *Treatise on the Auspicious Signs of Heaven and Earth*, a Han apocryphal text, *Extended Mystical Tally of the Classic of Filiality*, states that birds and beasts turn white to signify the virtuosity of the ruler. By contrast, when regular black crows sing, it often portends trouble in the form of bandits or robbery.[38] In the Northern and Southern Dynasties (386–589), the singing of black crows might also indicate that a man would be jailed because a woman was lying and gossiping about him.[39] Therefore, if one heard black crows singing, they should be warned about losing property and being gossiped about.

---

35 *Jinshu* 28.864 (五行志), 94.2435 (隐逸列傳).

36 *Tiandi ruixiang zhi* 18. 385.

37 *Hou Hanshu* 13.3273 (五行志).

38 *Tiandi ruixiang zhi* 18.389.

39 *Tiandi ruixiang zhi* 18.394.

Other birds' singing might predict different things. Vultures are also commonly observed birds in ancient China. The singing of vultures is interpreted differently, depending on the twelve earthly branches used for dates. On zi_1 子 days, the singing of vultures predicts the coming of bandits and thieves, but on yin_3 寅 days, it predicts the arrival of guests. Even within one day, as one day was divided into twelve dual hours, the singing of vultures at different double hours also predicts different things. For example, if they sing during the zi_1 子 double hour (23:00–1:00), this predicts illness. If the singing occurs during the yin_3 double hour (3:00–5:00), it predicts a visitation by guests. If it happens during the mao_4 卯 double hour (5:00–7:00), it indicates that there would be a lawsuit for the household.[40] This knowledge and these techniques are more useful for ordinary households to predict what would happen to them. Vultures, swallows, and sparrows are also commonly observed in urban and rural areas.

Many insects appear in the *Treatise on the Auspicious Signs of Heaven and Earth*, yet not all of them play significant roles in medieval Chinese zoomantic practices. Cicadas appear in many eulogies throughout the ages for their singing sounds, but do not draw much attention in terms of their zoomantic role. However, the *Treatise on the Auspicious Signs of Heaven and Earth* does mention that if flies assembled, it would indicate that calumny approached a ruler.[41] Many of the other fragments recorded in this encyclopedia mention the implications of these insects' appearances in the daily life. For example, Jing Fang's commentary on the *Book of Changes* claims that it is not a good sign if butterflies block a path. If flying butterflies surround troops, it indicates that there might be hidden enemies and the troops should be alarmed. Bees are also regarded as a warning of killing. During the Han, when General Jia Meng 賈萌 was persuaded to attack Wang Mang 王莽 (45 BCE–23 CE), his friend observed bees flying around his chariot and alerted him of this ominous sign. Later Jia Meng indeed was killed.[42] The *Master Embracing Simplicity* (*Baopuzi* 抱樸子) also warns that when troops marched with bees and horseflies flying around, there would be a sudden ambush and cruel battle because these insects indicated hidden enemies nearby.[43] The appearance of bees might also indicate illness, being jailed, robbery, or gossip gravely disturbing a family. If spiders construct their webs in the kitchen, it indicates that similarly unfortunate things would happen.[44] Locusts often appear as the destroyers of crops.

40 *Tiandi ruixiang zhi* 18.396–97.
41 *Tiandi ruixiang zhi* 18.418.
42 *Tiandi ruixiang zhi* 18.420.
43 *Tiandi ruixiang zhi* 18.420.
44 *Tiandi ruixiang zhi* 18.421.

Jing Fang's commentary on the *Book of Changes* claims that locusts eating grains means the administrators, who are paid in grain, are not the right choices. If the proper administrators were selected, the locust disaster would cease.[45] Again, this interpretation shows the correlative cosmology that could be traced back to the Han. Ants were also a bad omen, foreshadowing medical, social, and juridical troubles to a family. However, the discovery of earthworms was regarded as an auspicious sign.[46]

Fish, turtles, and crabs are different from birds and insects, though they are listed together with birds and insects in the medieval encyclopedia *Treatise on the Auspicious Signs of Heaven and Earth*. As I mentioned briefly in the previous section, flatfish were regarded as an auspicious sign for virtuous governance. However, it was a bad omen if fish came out from the water and endangered themselves. Jing Fang claimed that fish would appear in the sea if worthy people were selected to serve in the government, but it was an unfortunate sign if fish in the river swim against the stream and even ended up on paths.[47] Turtles were generally viewed as an auspicious animal. Han historian and writer Chu Shaosun 褚少孫 (fl. 32–7 BCE) suggested that the pronunciation of the character *gui* 龜 (turtle) was the same as *gui* 歸 (belonging to), which meant that one could collect property and became rich.[48] Many apocryphal texts claimed that divine turtles or black turtles granted diagrams or talismans to sage kings in the ancient past, which could be regarded as a sign of legitimation of their royal powers and authority. The legendary origins of the *Yellow River Diagram* (*Hetu* 河圖) and the *Writ of Luo River* (*Luoshu* 洛書) were connected with the turtle. A dark turtle granted a diagram to the Yellow Emperor. A talisman was granted to Yao 堯, an ancient sage king, according to the Han apocryphal text *Matched Prophecy and Realized Divination of the Classic of Documents*.[49] The dark turtle is also regarded as a divine turtle who was three hundred years old. In contrast to turtles, the crab is not an auspicious sign. According to Jing Fang's commentary on the *Book of Changes*, if crabs appeared in the court, there would be an armed coup d'état against the ruler. If crabs appeared amongst troops, it would indicate that the military campaign would fail.[50]

---

45 *Tiandi ruixiang zhi* 18.434.

46 *Tiandi ruixiang zhi* 18.435.

47 *Tiandi ruixiang zhi* 18.423.

48 *Tiandi ruixiang zhi* 18.425, which might cite from *Shiji* 128.68 (龜策列傳), which was written by Chu Shaosun. On Chu Shaosun, see David Knechtges and Tai-p'ing Chang, eds., *Ancient and Early Medieval Chinese Literature (Vol. I): A Reference Guide*, Part One (Leiden: Brill, 2010), 162.

49 *Tiandi ruixiang zhi* 18.426.

50 *Tiandi ruixiang zhi* 18.428–29.

The ancient lexicon *Erya* defines beasts as four-footed animals with skin covered in fur.[51] In the ancient zoomantic system, beasts included both mythical and real nonhuman animals. The mythical beast could be half animal half human, or half one animal half another animal, or even something with various body parts from different animals, such as the kirin (Qilin 麒麟). The basic idea was that different animals had their own natural behaviors. Abnormal or deviant behaviors delivered a prophecy. Jing Fang's commentary on the *Book of Changes* claimed that the king or lord would die within one year if wild animals entered villages with blood. Troops would invade households within one thousand days if wild animals entered local markets.[52] The *Book of One Hundred and Twenty Divinations* (*Bian nian zhan* 百廿占) claims that wild beasts' having intercourse with domestic beasts might prophesize a regime's collapse because of illicit sexual relations between a ruler and a woman from a foreign country.[53]

Many apocryphal texts, including the *Diagram of Auspicious Responses* (*Ruiying tu* 瑞應圖), regarded the kirin as a righteous beast.[54] The kirin had the body of a deer, a buffalo's tail, a wolf's roar, horse feet, and ox horns. To promote their political ideology, some Confucian apocryphal texts often claimed kirin would show up if a ruler practiced benevolent governance. Under the influence of Confucian political thought, many official histories recorded stories of the appearance of kirin, especially white ones. The elephant was also an auspicious animal. As the *Diagram of Auspicious Responses* states, the elephant would show up if righteous governance and teaching were extended in four directions.[55] The *Book of the Later Han Dynasty* claims that in 162 CE, frightened horses and unhinged elephants invaded the royal palace, which prophesized the disorder of the Han regime.[56]

In the Han apocryphal texts, if a horse exhibits some unusual changes, such as a horn growing on the head or giving birth to a baby pony with only three feet, it prophesizes political disorder in the court or even rebellion. Jing Fang's commentary on the *Book of Changes* also offers a list of interpretations concerning various strange body parts on horses. If a horse had three

---

51 *Erya* 10.321 (釋獸).

52 *Tiandi ruixiang zhi* 19.448–49.

53 *Tiandi ruixiang zhi* 19.449.

54 *Tiandi ruixiang zhi* 19.450. For its cultural importance as an auspicious animal in medieval China, see Sanping Chen, *Multicultural China in the Early Middle Ages* (Philadelphia: University of Pennsylvania Press, 2012), 51.

55 *Tiandi ruixiang zhi* 19.451.

56 *Hou Hanshu* 30.3345 (五行志).

eyes, it would indicate that the subjects controlled the ruler. If a horse had no eyes, it would indicate that the ruler would become ill for a long time. If a horse had no ears, it would prophesize that a ruler would eventually lose his throne. If a horse had no fur but feathers, it would indicate that there would be rebellions against the ruler.[57] Other domestic beasts, such as sheep and dogs, are subjected to related interpretations. If a sheep or dog was born in a wicked form, such as horns growing on the head, it would prophesize a grave future for a regime. Dogs behaving strangely also prophesizes health, legal, and social troubles in households, for dogs are familiar companions for many families at the local level.

Oxen were associated with agriculture because they were used in farming and transportation in ancient China. Endangering oxen would severely harm economic activities. Unusual behaviors of oxen also indicated a negative impact on economic activities. According to Jing Fang's commentary on the *Book of Changes*, if the government mobilized people for public projects and reduced their time in agricultural production, dire consequences would be manifested by the buffalo's having five feet.[58] If a buffalo had five feet, it meant that it lost the ability to serve in the labor force. If normal agricultural activities could not stay on track, insufficient grains would be produced and offered to the government, eventually undermining the regime. Nevertheless, most ancient regimes attempted to preserve their oxen herd. Other apocryphal texts indicated that political authority might wane and falter if oxen and horses roared in the palace.

Another interesting case on the relation between the behaviors of oxen and agriculture can be found in the *Anecdotes from the Jade Hall* (*Yutang xianhua* 玉堂閒話) by Wang Renyu 王仁裕 (880–956).[59] It claims that in the Five Dynasties (907–960), people often practiced verification with the behaviors of oxen on the first day of the first lunar month. If all oxen crouched, then the seedlings of the five grains might not grow well. If half of the oxen

57 *Tiandi ruixiang zhi* 19.461. For the study of the horse cult, see Meir Shahar, "The Chinese Cult of the Horse King, Divine Protector of Equines," in *Animals and Human Society in Asia: Historical, Cultural and Ethical Perspectives*, eds. Rotem Kowner, Guy Bar-Oz, Michal Biran, Meir Shahar, Gideon Shelach (Dordrecht: Springer, 2019), 355–90.

58 *Tiandi ruixiang zhi* 19.463. As I discussed in Chapter 6, local villagers in many places in China and India believed that the were-tigers would have been exposed by their five-foot prints.

59 Zhao Xianhuan, *Classical Chinese Supernatural Fiction: A Morphological History* (Lewiston: E. Mellen Press, 2005), 83–84.

crouched and half of them stood up, the yearly harvest would be average. If all oxen stood up, there would be a tremendous harvest of the five grains.[60]

There is a very long history of rearing pigs in China. Pigs were regarded as water beasts in many ancient texts for their approach to water, including rivers, in the humid and hot monsoon climate in China. While this behavior was observed, it was used for predicting the coming of rainstorms. Then, in ancient China a cult of the pig as the water beast has been gradually developed. This cult is mixed with the old tradition of divination by cloud vapors, which can be found in the Tang divination text *Book on Interpreting Rain* (*Xiangyu shu* 相雨書) by Huang Zifa 黃子發. Eventually, it developed into a cult of the Great Prime Marshall Tianpeng (Tianpeng yuanshuai 天蓬元帥) in the heavenly court, a deity believed to be in charge of rain.[61] However, this function of divination does not appear in the *Treatise on the Auspicious Signs of Heaven and Earth.* This text focuses on the appearance of undomesticated hounds (*shi* 豕) in places where they were not supposed to appear, such as the official court or villages.

Villagers and residents could encounter wild beasts while traveling around; hermits also encountered them upon entering the mountains and forests. The tiger was viewed as the king of mountain beasts for it occupied the top of the food chain in the Chinese ecological system. The *Diagram of Auspicious Responses* suggests that white tigers are righteous beasts who would never harm human beings, which indicated the virtuous governance of the ruler.[62] The *Spring and Autumn Explanatory Kong Chart* also claims that the mandate of heaven would be restored if a white tiger appeared in the court, but a regime would come to an end if one was found in the field.[63] The *Brief [History] of Wei* (*Weilüe* 魏略), a historical work compiled between 239 and 265 CE, states that when Emperor Wendi 魏文帝 (r. 220–226) or Cao Pi 曹丕 (187–226) was about to be enthroned, local kingdoms reported the appearance of white tigers twenty-seven times.[64] It was commonly believed that tigers would not actively attack humans. If they did so, it was regarded as a sign of the victim's corrupted morality and destructive behaviors.[65] Jing Fang's commentary on

60 *Leishuo jiaozhu* 54.1613.

61 Min Xiangpeng and Huaiyu Chen, "Animals, Divination, and Climate: An Environmental Note on the Cult of the Pig in Ancient China," *ISLE: Interdisciplinary Studies in Literature and Environment* (2021), https://doi.org/10.1093/isle/isaa195. See Chapter 3 of this book.

62 *Tiandi ruixiang zhi* 19.478.

63 *Tiandi ruixiang zhi* 19.479.

64 *Tiandi ruixiang zhi* 19.479.

65 For the cultural meaning of the white tiger, see Hou Ching-lang, "The Chinese Belief in Baleful Stars," in *Facets of Taoism: Essays in Chinese Religion*, eds. Holmes Welch and Anna Seidel (New Haven and London: Yale University Press, 1979), 209–19.

the *Book of Changes* states that tigers eat humans if the ruler is greedy and the subjects are cruel and violent.[66]

Although white wolves are viewed as auspicious signs of virtuous governance in the *Diagram of Auspicious Responses*, it was not a good sign if wild wolves entered urban areas.[67] In *Master Lü's Spring and Autumn Annals*, a wolf entering the state was a sign of deviance and portended damage to the state.[68] Jing Fang's commentary on the *Book of Changes* also states that the roaring of wolves in a village indicated property loss, possibly because of a violent war against the local community. If dark wolves appeared in the villages, hungry residents might eat each other.[69] Interestingly, the *Treatise on the Auspicious Signs of Heaven and Earth* did not associate wolves with nomadic tribes as the danger of invasion was from the northern frontiers. The reason was unknown. However, it might be due to the taboo the author observed that the royal house of the Tang regime had kinship connection with the Xianbei 鮮卑 tribe in the north.

The reindeer and fox are connected with licentious behaviors of women in medieval Chinese divination. The unusual appearance of either of these beasts in a human residence indicates that something wrong happens within the household, damaging the family. The fox is viewed as a particularly deviant animal who would bring illness, death, and livestock loss.[70] The *Treatise on the Auspicious Signs of Heaven and Earth* also mentions mythical animals such as the *baize* 白澤, the *jiaoduan* 角端, and the *zhuru* 朱儒 that can be identified with real nonhuman beasts. Many of these beasts are deviant or strange.[71]

Rodents were in general regarded as ominous signs. As the *Diagram of Auspicious Responses* claims, if a white mouse is one *chi* 尺 (foot) long and has four red feet, it is a deviant mouse. The collection of strange writings *Garden of Marvels* (*Yiyuan* 異苑) records a story about an official Gao Huiqing 高惠清 (fl. fifth century CE). During 397–401, a group of mice linked together with heads and tails, falling from Gao's house beam down to the ground, and Gao became ill for several days.[72] Jing Fang's commentary on the *Book of Changes*

66 *Tiandi ruixiang zhi* 19.480.

67 *Tiandi ruixiang zhi* 19.480.

68 *Tiandi ruixiang zhi* 19.480–81.

69 *Tiandi ruixiang zhi* 19.481.

70 *Tiandi ruixiang zhi* 19.491–94. For the comprehensive study on the fox, see Kang Xiaofei, *The Cult of the Fox: Power, Gender, and Popular Religion in Late Imperial and Modern China* (New York: Columbia University Press, 2006).

71 *Tiandi ruixiang zhi* 19.501–2.

72 *Tiandi ruixiang zhi* 19.510.

also indicated that if a group of mice assembled in the royal hall, the ruler might die. If a group of mice entered the village and roamed around, local residents might be in trouble. If a bunch of mice walked out and consumed five grains, it prophesized worthy subjects distanced from the ruler, and deviant subjects not only pleasing to the ruler, but also confounding virtuous governance.[73] The deviant behaviors of mice also indicated other troubles such as bandits, robbery, lawsuits, and illness to ordinary households. Surprisingly, no plague brought by rodents is mentioned in Sa Shenzhen's *Treatise on the Auspicious Signs of Heaven and Earth.*

Lastly, I will briefly discuss dragons and snakes. The dragon was not a real nonhuman animal but a mythical animal. It was similar to the kirin in terms of its association with royal authority and legitimation.[74] For instance, the *Diagram of Auspicious Responses* states that the yellow dragon is the oldest among five kinds of dragons, and its appearance indicates the rise of a new king.[75] Other apocryphal texts such as the *Extended Mystical Tally of the Classic of Filiality* endorse this idea, but add that the king should have virtue.[76] Some apocryphal texts claim that deviant dragons and dark dragons are bad omens, foreshadowing rebellion, illness, and death.[77] Some apocryphal texts indicate that a snake entering into the capital city might lead to the death of the king's mother. Jing Fang's commentary on the *Book of Changes* claims that war will occur within three years if a snake enters a village.[78] *The Classic of Mountains and Seas* states that a wicked snake with six feet and four wings portends great drought under heaven, which will damage agricultural production and lead to rebellion and political chaos.[79] Snakes were not domestic animals and their abnormal behaviors, such as appearing in wicked forms or entering human spaces, often prophesized terrible things, such as the lord losing his mother. It is noteworthy that snakes might be associated with women for their ill-fated lives; other creatures did not have such an association.[80] It seems that gender does not play many significant roles in ancient Chinese zoomantic practices, though occasionally, women were indeed explicitly mentioned.

---

73 *Tiandi ruixiang zhi* 19.511.

74 Jean-Pierre Diény, *Le symbolisme du dragon dans la Chine antique* (Paris: Collège de France, Institut des hautes études chinoises, 1987); J. Keith Wilson, "Powerful Form and Potent Symbol: The Dragon in Asia," *The Bulletin of the Cleveland Museum of Art* 77, no. 8 (1990): 286–323.

75 *Tiandi ruixiang zhi* 19.517.

76 *Tiandi ruixiang zhi* 19.517.

77 *Tiandi ruixiang zhi* 19.518–19.

78 *Tiandi ruixiang zhi* 19.520.

79 *Tiandi ruixiang zhi* 19.522.

80 *Tiandi ruixiang zhi* 19.522–26.

Chapter 5

THE CHANGING IMAGES OF ZODIAC ANIMALS IN MEDIEVAL CHINESE BUDDHIST LITERATURE

Introduction

In the medieval period, specifically from the fifth to eighth centuries in the Chinese historical context, the Buddhist community produced a voluminous literature, which became an invaluable resource for understanding the roles of animals in medieval religious life. This Buddhist literature not only included numerous translations rendered from Indic and Central Asian languages such as Gāndhārī and Tocharian, or which many original texts were lost, but also commentaries and collections of magical stories. In these texts, there were encounters and adaptations between Buddhist ideas and Chinese indigenous thoughts. Unlike many other world and local religions, Buddhism developed a sophisticated cosmology in which animals occupied one of the six realms for the birth, death, and rebirth of all lives or sentient beings in Buddhist terms. The Buddhist idea of reincarnation created a blurry boundary between humans and animals. In other words, in Buddhist literature, humans may be reborn as animals and vice versa, which also applies to gods and spirits. Drawing upon sources from medieval Chinese Buddhist literature centered on the *Mahāvaipulya-mahāsaṃnipāta-Sūtra* (hereafter abbreviated to *MMS*),[1] this chapter examines how twelve animals first appeared as compassionate

1 Chinese title: *Da fangdeng daji jing* 大方等大集經. There are various translations of many sections of this huge collection in the Chinese canonical Buddhist literature. For a comprehensive ethnographic study on this message in Japanese scholarship, see Minakata Kumagusu 南方熊楠, "Jūnishi kō 十二支考 (A Study on Twelve Terrestrial Branches)," in *Minakata Kumagusu zenshū* 南方熊楠全集 (The Complete Works of Minakata Kumagusu), vol. 1 (Tokyo: Heibonsha, 1971), 5–610; this is a collection of a series of essays he published in a journal entitled *Taiyō*, during 1915–1924, which was later reissued as a monograph. In 1919, he also published an article on the twelve animals, see his "Shishin to jūnijū nitsuite 四神と十二獣について (On Four Deities and Twelve Beasts)," *Jinreigaku zashi* (人類學雜誌) 34, no. 8 (1919), reprinted

deities and then as demonic spirits in these sources and how Buddhist and Chinese intellectual traditions shaped this transformation.

The *MMS* is a sizeable collective scripture of early Mahāyāna Buddhist texts which did not survive in its entirety in any language. Some fragments in Sanskrit have been identified as parts of this significant scripture. Among numerous Sanskrit manuscripts found in Gilgit, Khotan, and other places in Central Asia, some fragments of the Sanskrit texts have been identified as those of the *Ratnaketudhāraṇī*,[2] which is equivalent to the ninth chapter (Baochuang fen 寶幢分) of the Chinese *MMS*. There is also a separate Chinese translation *Baoxing Tuoluoni Jing* 寶星陀羅尼經 translated by Prabhāmitra in the Tang Dynasty and the lengths is about twice of the ninth chapter of the Chinese *MMS*. Although there is no complete equivalent text in Sanskrit, the Chinese and Tibetan translations that perhaps represent

in *Minakata Kumagusu zenshū*, 2 (1971): 147–58. Many Chinese scholars such as Li Xueqin 李學勤, Li Ling 李零, Liu Guozhong 劉國忠, Cai Hongsheng 蔡鴻生, Li Shuhui 李樹輝, and Wu Yucheng 吳裕成 have studied the origins of twelve zodiac animals and their spread in Asia based on both traditional sources and excavated documents, including the bamboo slips from Fangmatan 放馬灘 in Gansu Province and Yunmeng 雲夢 in Hubei Province. I will not exhaust listing all articles here. However, it should note that Liu Guozhong also briefly mentioned twelve zodiac animals and thirty-six birds; see his "Shi lun shi'er shengxiao yu sanshiliu qin 試論十二生肖與三十六禽 (A Preliminary Study on Twelve Zodiac Animals and Thirty-Six Birds)," *Qinghua daxue xuebao* 清華大學學報 14, no. 1 (1999): 12–15; Cai Hongsheng, *Tangdai jiuxinghu yu Tujue wenhua* 唐代九姓胡與突厥文化 (The Nine-surname Sogdians and Turkic Culture in the Tang Dynasty) (Beijing: Zhonghua shuju, 1998), 164–85. Many scholars also noticed the twelve zodiac animals on the mural paintings of the Tang tombs; see Zhang Yun 張蘊, "Xi'an diqu Sui Tang muzhi wenshi zhong de shi'er shengxiao tu'an 西安地區隋唐墓誌紋飾中的十二生肖圖案 (The Twelve-Zodiac Animal Pictures in the Patterns decorated the Tombstones in the Sui and Tang Tombs in the Xi'an Region)," *Tang Yanjiu* 唐研究 8 (2002): 395–432.

2 For the Gilgit Manuscripts, see Nalinaksha Dutt, *Gilgit Manuscripts*, vol. 4 (Delhi: Sri Satguru Publications, Delhi 1984, reprinted, originally published in 1939); Raghu Vira and Lokesh Chandra, *Gilgit Buddhist Manuscripts* (New Delhi: International Academy of Indian Culture, 1974); Zuiryu Nakamura 中村瑞隆, "Hôshô Daraniki kyô dankan 寶星陀羅尼經斷簡 (Gilgit Manuscript of the *Mahāsannipātaratnaketusūtra*, kept in the National Archives, Katmandu)," *Hokekyô Bunka Kenkyû* 法華文化研究 no. 1 (1975): 13–45; Yenshu Kurumiya, *Ratnaketuparivarta*, Sanskrit Text (Kyoto: Heirakuji Shoten, 1978). Many Sanskrit fragments have been found in the manuscripts from Central Asia in many collections across the world; see A. F. Rudolf Hoernle, *Manuscript Remains of Buddhist Literature Found in Eastern Turkestan* (Oxford: Clarendon Press, 1916), 121–25; Seishi Karashima 辛嶋靜志, "Four Sanskrit Fragments of the *Ratnaketuparivarta* in the Stein Collection," in *The British Library Sanskrit Fragment: Buddhist manuscripts from Central Asia*, vol. 1, eds. by Seishi Karashima and Klaus Wille (Tokyo: The International Research Institute for Advanced Buddhology,

substantial portions, have survived in numerous scrolls and manuscripts. In the Tibetan canon, there is no complete equivalent of the MMS either. A number of texts in Kangyur have been identified to correspond some sections of the Chinese version. For example, the Tibetan text *'phags pa de bzhin gshegs pa'I snying rje chen po nges par bstan pa* (D 147) translated by Indian master Śīlendrabodhi and Tibetan master Ye shes sde in the eighth century is the equivalent to the first two sections of the Chinese *MMS*. There are also some smaller texts (D 91, 138, 148, 152, 169, 175, 230, 257, etc.) in the Tibetan canon that correspond to several sections of the Chinese MMS.[3]

This chapter will focus on only a small part of the Chinese translation by Dharmakśema 曇無讖 (385–433) in 414–426. In chapter 23 of the tenth scripture, the *Xukongmu fen* 虛空目分 (*Śunyācakṣus-) of the *MMS*,[4] twelve animals

Soka University, 2006), 177–89; Seishi Karashima, "The Sanskrit Fragments Or.15010 in the Hoernle Collection," in *The British Library Sanskrit Fragments*, vol. 2, eds. by Seishi Karashima and Klaus Wille (Tokyo: International Research Institute for Advanced Buddhology, Soka University, 2009), 335–550; Jens-Uwe Hartmann and Chanwit Tudkeao, "Three Sanskrit Fragments of the *Ratnaketuparivarta*," in *The British Library Sanskrit Fragments*, vol. 2, eds. by Seishi Karashima and Klaus Wille (Tokyo: International Research Institute for Advanced Buddhology, Soka University, 2009), 589–96; Chanwit Tudkeao, "Three Sanskrit Fragments of the *Ratnaketuparivarta*," in *The British Library Sanskrit Fragments*, vol. 3, eds. by Seishi Karashima and Klaus Wille (Tokyo: International Research Institute for Advanced Buddhology, Soka University, 2015), 589–91; for some fragments found in Xinjiang, see Duan Qing 段晴 and Zhang Zhiqing 張志清 eds., *Zhongguo guojia tushuguan cang xiyu wenshu: Fanwen, quluwen juan* 中國國家圖書館藏西域文書：梵文、佉盧文卷 (The Manuscripts from the Western Regions preserved at the National Library of China: the Sanskrit and Kharoṣṭhī Volume) (Shanghai: Zhongxi shuju, 2013), 41–119; for a general survey, see Saerji 薩爾吉, *Dafangdeng daji jing yanjiu* 大方等大集經研究 (A Study of the *Mahāvaipulya-mahāsaṃnipāta-Sūtra*) (Shanghai: Zhongxi shuju, 2019), 4–11.

3 Saerji, *Dafangdeng daji jing yanjiu*, 52–60. D 356 has been studied and translated by Jan Nattier, *Once Upon a Future Time: Studies in a Buddhist Prophecy of Decline* (Berkeley: Asian Humanities Press, 1991), ch. 8. Some manuscripts in old Tibetan also have been identified in the Dunhuang manuscripts, see Jacob Dalton and Sam van Schaik, *Tibetan Tantric Manuscripts from Dunhuang* (Leiden: Brill, 2006), 25–29.

4 Seishi Karashima suggested that "the description concerning the Mahāsāṃghikas appears at the end, and it is expressed in a positive way, while other schools are described negatively. We may assume that the composer of this vaitulya-scripture belonged to the school of the Mahāsāṃghikas"; see his article "Who Composed the Mahāyāna Scriptures?—The Mahāsāṃghikas and Vaitulya Scriptures*," *Annual Report of The International Research Institute for Advanced Buddhology at Soka University for the Academic Year 2014* 18, (2015), 139. This scripture also mentions some interesting place names in Central Asia, such as Kucha, Khotan, and Gandhara, which illustrated its close connection with the Buddhist culture in Central Asia.

appeared as the reincarnations of Buddhist Bodhisattvas in a calendrical circle of twelve days. They would travel to the realm of animals to enlighten those sentient beings in suffering. This passage frequently appeared in numerous commentaries by the later generations of Buddhist masters who attempted to interpret this passage to the Chinese audience by incorporating indigenous Chinese ideas, such as the Yin-Yang and Five-phase theories. Nevertheless, in these interpretations in the following centuries, the twelve animals were transformed into demonic spirits who were said to corrupt the morality of Buddhist practitioners, thus losing their original identity as Bodhisattvas who attained enlightenment but remained to enlighten other sentient beings.

Twelve Animals as Reincarnations of Bodhisattvas in the *MMS*

Chapter 23 of the *MMS* describes a Buddhist cosmology whereby many Bodhisattvas appeared like animals. It is said that there are many Bodhisattvas in the world. They could appear in the form of any sentient being in the six realms, such as gods, demigods, humans, animals, hungry ghosts, and demons, and travel around the realm of Jambudvīpa (worldly continent), teaching and transforming sentient beings. It claimed that it would not be difficult to teach and convert the beings in the sentient realms of gods and humans, but it would be uneasy to teach and convert the sentient beings in the realm of animals. Then, the text tells how the Bodhisattvas appeared in various forms of the animals in order to teach these animals Buddhism.

Then, the passage in chapter 23 of the *MMS* tells that there are four seas and four mountains surrounding this realm of Jambudvīpa. In the south, there is a mountain of lapis lazuli, which is called the "tide." On this mountain, as the reincarnations of Bodhisattvas, three animals, namely a poisonous snake, a horse, and a sheep, resided in three caves (with three names called various forms or rūpa 種種色, no death 無死, and well-established 善住 or supratiṣṭhita in Sanskrit), respectively. They all cultivated the compassion of the listeners (shengwenci 聲聞慈) of Buddhism. A tree god (wusheng 無勝), a guardian (Rākṣasī, whose name is kind actions 善行), and five hundred attendants served as their companions. The three animals received offerings as Bodhisattvas. In the west, there is a mountain of rock crystal. There are three caves (*shangse* 上色, *shiyuan* 誓願 or vows, and *fachuan* 法床 or Dharma bed) for the Bodhisattvas who appeared as three animals: a monkey, a rooster, and a dog. They also help cultivate the compassion of the listeners of Buddhism. A fire god, a guardian (Rākṣasī, whose name is seeing by eyes 眼見), and five hundred attendants serve as their companions. There is a mountain of silver in the north (whose name is *putiyue* 菩提月,

the Bodhi Moon), with three caves (diamond 金剛, virtues and merits of incense 香功德, and high virtues and merits 高功德) for the reincarnation of three Bodhisattvas, appearing as three animals: a pig, a mouse, and an ox, who help cultivate the compassion of the listeners of Buddhism. A wind god on the mountain (whose name is moving wind), a guardian (Rākṣasī, whose name is heavenly protection 天護), and five hundred attendants serve as their companions. There is a mountain of gold in the east, with three caves (bright star 明星, pure way 淨道, and joy and happiness 喜樂) for three animals as the reincarnations of Bodhisattvas: a lion, a rabbit, and a *nāga*, who, again, cultivate the compassion of the listeners of Buddhism. A water god (whose name is the Heaven of Water 水天), a guardian (Rākṣasī, whose name is cultivating the shame 修慚愧), and five hundred attendants serve as their companions. In total, there are twelve animals. They travel across the realm of Jambudvīpa from day to night, receiving venerations from sentient beings in the six realms.

The *MMS* states that these animals have accomplished their merits and virtues. In the place of the Buddhas, they would make profound and serious vows of taking turns to travel around for the purpose of teaching and transforming sentient beings each day in a circle of twelve days. For example, on the first day of the seventh month, the mouse begins to travel around to teach and transform all sentient beings who are reborn in the body form of the mouse, depriving their evil deeds and advising them to cultivate good deeds. Then the next day, another animal would come and replace the mouse. On the thirteenth day, the mouse starts the journey again. In the same way, this scheduling also runs for twelve months and even twelve years.[5]

Several striking points should be noted here. First of all, this passage tells us that each animal was a Bodhisattva and all twelve animals, as the reincarnations of the Bodhisattvas, lived in places beyond the realm of Jambudvīpa, which means that they are enlightened beings but remain to teach sentient beings. In other words, they accomplished their merits and virtues and became Bodhisattvas, being revered by sentient beings, including heavenly and human beings. Since they were Bodhisattvas, they could come out of their dwelling caves and appear in the form of animals in this worldly realm for teaching and transforming sentient beings. They are responsible for teaching and transforming those sentient beings who are reborn as animals. They follow a calendrical order of twelve days, twelve months, and twelve years and come out by the turns of twelve animals. Nevertheless, as a typical

5 *Dafangdeng daji jing*, juan 23, *T.* 13, no. 397: 167b–168b.

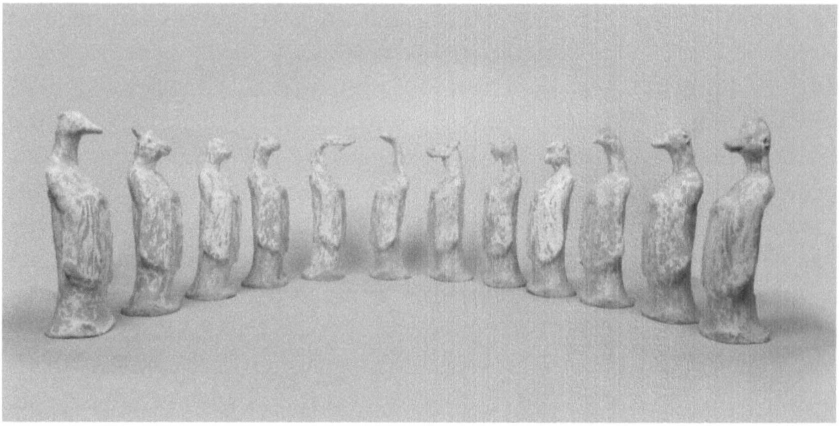

Figure 5.1: A set of figurines of twelve zodiac animals from the Metropolitan Museum of Art (Accession Number: 2000.662.7a–l). Eighth century. As part of The Met's Open Access program, the data is available for unrestricted commercial and noncommercial use without permission or fee.

Mahāyāna idea, the Bodhisattvas could transform themselves into whichever body form of six categories of sentient beings, including beasts, as long as they wish, for the sake of teaching and transforming other sentient beings, to cultivate merits and virtues.

Moreover, the twelve animals listed in the passage of the *MMS* reflect a strong Indo-centric worldview. It starts from the mouse and is followed by the ox, lion, rabbit, *nāga*, snake, horse, sheep, monkey, rooster, dog, and pig, which is different from the Chinese list of twelve zodiac animals that includes a tiger, rather than a lion (Figure 5.1).[6] Second, it also touches on the ancient Indic concept of the material world constructed by four great elements. According to the passage in the *MMS*, these animals are said to reside in the caves of four mountains where eight female deities and their attendants served as companions. Among these eight female deities, four of them are named after the tree, fire, wind, and water, symbolizing the four great elements (Skt: *cattāro mahābhūta*; Pali: *cattāro mahābhūtāni*) in ancient Indian cosmology. Although it was usually listed as earth, water, fire, and air or wind in the Pali

6 Zhang Xing and Chen Huaiyu, "From Lion to Tiger: The Buddhist Changing Images of Apex Predators in Trans-Asian Contexts," in Rotem Rosen, Michal Biran, Meir Shahar, and Gideon Shelach eds., *Animals and Human Society in Asia: Historical and Ethical Perspectives* (Basingstoke: Palgrave Macmillan, 2019), 331–53.

texts and the Chinese translations of many early Buddhist texts,[7] here chapter 23 of the Chinese MMS listed as earth, fire wind, and water, which is slightly different. Third, this passage mentioned four mountains in four cardinal directions surrounding the Jambudvīpa world, which started from the south, followed by the west, the north, and the east. In the south, the three animals included a snake, a horse, and a sheep. In the west, the three animals included a monkey, a rooster, and a dog. In the north, the three animals were a pig, a mouse, and an ox. Whereas in the east, the three animals were a lion, a rabbit, and a *nāga*. Interestingly, from the perspective of medieval Chinese Buddhists, this arrangement seems to correspond with the notion that the Indian subcontinent was the center of the world in the Indo-centric worldview.

The most cited modern scholarly edition of the *MMS* in the *Newly compiled Buddhist Canon in the Taishō period* (Taishō Shinshū Daizōkyō 大正新脩大藏經) published in Japan was based on the Korean edition carved in the tenth century,[8] which does not reflect the original structure and vocabulary of

---

7 The normal order of these four elements usually lists as earth element (pruṭhavī-dhātu), water element (āpa-dhātu), fire element (teja-dhātu), air (or wind) element (vāyu-dhātu). They constituted the form (rūpa) of material realm. See *The Middle Length Discourses of the Buddha: A Translation of the Majjhima Nikaya*, translated by Bhikkhu Ñāṇamoli and Bhikkhu Bodhi (Boston: Wisdom Publications, 1995), 278. For a brief introduction, also see Hajime Nakamura, *Buddhist Cosmology: Philosophy and Origins* (Tokyo: Kosei Publishing Co., 1997), 21. In the Chinese translations of some early Buddhist texts, such as *Kevaddha-sutta* (Jiangu jing 堅固經, *Dīrgha Āgama* or *Chang ahan jing* 長阿含經, juan 16), Brahmanimantanika-sutta (Fantian qingfo jing 梵天請佛經, *Madhyama Āgama* or Zhong ahan jing 中阿含經, juan 19), *Samyuktāgama* (Za ahan jing 雜阿含經), *Ekottarika-āgama* (Zengyi ahanjing 增一阿含經), and *Lalitavistara* (Puyao jing 普曜經) the order is the original: earth, water, fire, and air/wind.

8 The stone scriptures preserved in the Fangshan area carved in 1063 preserved an early version of the MMS. The corresponding passage appeared in the stone tablet numbered in Chinese character "you 有," located in the no. 212 scripture in the seventh cave of the Fangshan area. The order of the four mountains started from the south, and then the west, the north, and the east. The manuscripts from Dunhuang indeed preserved some fragments of the Chinese MSS. For example, chapter 23 appeared in the fragments S. 1261 and Bei 8714, but these fragments did not preserve the section on the current passage here I focused on; see Huang Yongwu 黃永武 ed. by *Dunhuang baozang* 敦煌寶藏 (Treasures from Dunhuang) (Taipei: Xinwenfeng chuban gongsi, 1986), vol. 9, 475ff and vol. 111, 260ff. Another fragment P. 2108 (*Dunhuang baozang*, vol. 114, 412ff) preserved some sections from chapter 24 and the fragment S. 582 (*Dunhuang baozang*, vol. 4, 621ff) has sections from chapter 25. The fragment Ch. 5506 (T M 48 D 56) in the Turfan manuscripts is also from chapter 23 of the MMS, but it does not preserve the corresponding section of this passage on twelve animals; see G. Schmidt and Thomas Thilo, *Katalog chinesischer buddhistischer Textfragmente I*

the text that was translated in the early fifth century. In fact, in the editions printed in China in the following centuries, the order of the four directions beyond the Jambudvīpa realm was modified by the Chinese compilers by starting from the east, followed by the south, the west, and the north. This obvious Chinese Buddhist modification reflected the Chinese Buddhist transformation from the Indo-centric worldview to the Sino-centric worldview. The Chinese Buddhist narratives often regarded the east, where China is located, as the primary direction, which is also the direction where the sun rises.

Further, the MMS tells that the four mountains beyond the Jambudvīpa realm refer to the mountains of lapis lazuli in the south, crystal in the west, silver in the north, and gold in the east, which were the first four jewels out of seven treasures in Buddhist culture.[9] Again, the order of listing these four jewels is worth noting. Different cultures might value different jewels due to various historical, material, and cultural reasons, or even just because of accessibility and aesthetics. Perhaps the lapis lazuli was valued in South Asia, so the lapis lazuli world was listed first. Yet, in ancient China, gold was valued more than many other treasures, for its importance in serving as currency, rewards, and decorative materials. So, the gold in the east was listed first in the passage of chapter 23 of the Chinese *MMS*. We can also look at other Chinese Buddhist texts. The four jewels were popular names that also appeared in some Chinese translations of early Buddhist texts. Most of the lists of four jewels began from the east with the gold. For example, the list beginning with the gold mountain in the east appeared in the *Scripture spoken by the Buddha on Nirvana* (Foshuo nihuan jing 佛說泥洹經) translated by Bai Fazu 白法祖 in the western Jin Dynasty, the *Scripture on Lokadhātu* (Da Loutan jing 大樓炭經) translated by Fali 法立 and Faju 法炬

(BTT VI) (Berlin: Wiley-VCH Verlag, 1971), 141, plate 13/15. Therefore, there is no way to know if the manuscript of the MMS from the medieval period preserved an earlier version of this passage or not.

9 Many Buddhist texts list seven treasures including suvarṇa (金, gold), rūpya (銀, silver), vaiḍūrya (琉璃, aquamarine / lapis lazuli), sphaṭika (頗梨, crystal), musāragalva (硨磲, conch shell / white coral), lohita-muktikā (赤珠, ruby / red pearl), and aśmagarbha (瑪瑙, emerald / carnelian). See Paul J. Griffiths trans., *The Realm of Awakening: A Translation and Study of the Tenth Chapter of Asaṅga's Mahāyānasaṅgraha* (Oxford: Oxford University Press, 1989), 209; Chün-fang Yü, *Chinese Buddhism: A Thematic History* (Honolulu: University of Hawai'i Press, 2020), 204. Although I used lapis lazuli to translate the Chinese word "liuli" in this context, "liuli" could be more sophisticated in medieval Chinese Buddhist literature. For a comprehensive study on this word "liuli" in medieval China, see Yu Xin, "*Liuli* in Buddhist Rituals and Art in Medieval China," *Hualin International Journal of Buddhist Studies* 1.2 (2018): 231–68.

in the western Jin Dynasty, the *Makhādeva-sutta* (Da tian nailin jing 大天奈林經) (of the *Majjhima-nikāya*) translated by Gautamasaṃghadeva 瞿曇僧伽提婆 in the Eastern Jin Dynasty, and the *Mahāparinibbāṇa Sutta* (Youxing jing 遊行經) (of the Dīgha Nikāya) translated by Buddhayaśas 佛陀耶舍 and Zhu Fonian 竺佛念 in the Eastern Jin Dynasty. One fragment of the Buddhist manuscript from Dunhuang even claimed that Mount Sumeru was also called Mount Kunlun, which was constituted by four jewels. Its eastern side was gold, its western side was silver, its southern side was lapis lazuli, and its northern side was crystal.[10] The orderly arrangement of these four jewels and their relevant directions seems to be another piece of evidence for transforming the Indo-centric worldview to the Sino-centric worldview.

More importantly, in reading the Chinese versions of the *MMS*, the changing cardinal directions for the mountains where animals dwell should be understood by looking into its historical root in the political culture of ancient China.[11] In ancient Chinese texts, such as the turtle bone inscriptions and the *Book of Documents* (Shangshu 尚書), the four directions were commonly ordered as the east, the south, the west, and finally the north. But in some other texts, such as the *Zuo Commentary on the Spring and Autumn* (Zuozhuan 左傳) and the *Record of Rites* (Liji 禮記), the positions of south and west were switched, that is, east, the west, the south, and the north. However, both make the east the primary and foremost position, which might reflect a Chinese Buddhist concept.

From the cultural geography perspective in the Buddhist tradition, early Buddhist literature often introduced and explained four continents (*catvaro dipah*) of the world by starting from the continent in the south, which appeared as Jambudvipa. This is not a random appearance but reflects

10 T. 85, no. 2899, 1427b. The title is the Aśvaghoṣa chapter of the *Lotus Sutra*. But it only appeared in the Dunhuang manuscripts, not in other translations of the *Lotus Sutra*. So it might be a popular apocryphal in medieval China and to some extent reflected the medieval worldview of Chinese Buddhists.

11 For a study of the four directions in ancient Chinese political thought, see Aihe Wang, *Cosmology and Political Culture in Early China* (Cambridge: Cambridge University Press, 2000), chapter two: "Sifang and the center: The Cosmology of Ruling China," 23–74. However, she does not elaborate on the order of the four directions. Also see Laurent Sagart, "The Chinese Names of the Four Directions," *Journal of American Oriental Society* 124, no. 1 (2004): 69–76. Cai Zhemao 蔡哲茂, "Jiaguwen sifang feng ming zaitan 甲骨文四方風名再探 (Revisiting the names of four-directional winds in the Oracle Bone Inscriptions)," *Jin Xiangheng jiaoshou shishi zhounian jinian lunwenji* 金祥恆教授逝世週年紀念論文集 (Collected Papers in memory of the one-year anniversary of the Passing of Professor Jin Xiangheng) (Taipei: Jin Xiangheng jiaoshou shishi zhounian jinian lunwenji bianji xiaozu, 1990), 123–52; Zhou Xiaolu 周曉陸, "Shi Dongxinanbeizhong: jianshuo ziwu 釋東西南北中：兼說子午 (Notes on East, West,

the cosmological worldview of people in ancient South Asia, where Buddhism began to flourish. Hence, unsurprisingly this cosmological worldview was built upon South Asian centrism. As Akira Sadakata 定方晟 pointed out, in the early Buddhism, or using his words, pre-Mahāyāna Buddhism, the four continents spread out by surrounding the Sumeru Mountain and one of the four continents, Jambudvipa, was modeled the subcontinent of India. So, he highlights that the features of the Jambudvipa, as illustrated in the early Buddhist literature, reflected the world geographical knowledge of ancient Indian people.[12] The Buddhist text *Mahāratnakūṭa-sūtra* 大寶積經 claimed that the *Jambudvipa* was unique because it was much larger than the other three continents, with one hundred million yojana in width. It has four times more cities than the other three continents. Moreover, these cities were very magnificent, peaceful, joyful, and prosperous. Most importantly, the Buddha resided in this continent.[13] Ancient Buddhism seems to attempt to glorify this continent that modeled the subcontinent of South Asia, where the Buddha was born, grew up, reached the enlightenment, and preached his Dharma.[14]

South, North, and Center: Also on Zi Wu)," *Nanjing daxue xuebao* 南京大學學報, no. 3 (1996): 70–76; Zhang Dexin 張德鑫, "Fangweici de wenhua kaocha 方位詞的文化考察 (A Cultural Perspective on the Vocabulary of Cardinal Directions)," *Shijie hanyu jiaoxue* 世界漢語教學 37, no. 3 (1996): 62–73; Chu Zexiang 儲澤祥, "Hanyu kongjian fangwei duanyu de lishi yanbian de jige tedian 漢語空間方位短語的歷史演變的幾個特點 (Several Features on the Historical Development of the Phases on Cardinal Directions and Spaces in Chinese Language)," *Gu hanyu yanjiu* 古漢語研究 30, no. 1 (1996): 57–61; Li Qiwen 李啟文, "Lun guren shi ruhe kandai fangwei de 論古人是如何看待方位的 (How did ancient People View the Cardinal Directions)," *Xueshu yanjiu* 學術研究, no. 2 (1999): 72–75.

12 Akira Sadakata, *Buddhist Cosmology: Philosophy and Origins*, trans. by Gaynor Sekimori, with a foreword by Hajime Nakamura (Tokyo: Kōsei Publishing Co., 1997), 31.

13 *T.* 11, no. 310: 141b.

14 It seems not surprising that many people in the ancient times viewed the geographical region where they lived and developed their culture as the center of the world. In ancient Tibet, Tibetans also viewed Tibet as the center of the world; see Dan Martin, "Tibet at the Center—A Historical Study of Some Tibetan Geographical Conceptions Based on Two Types of Country-lists Found in Bon Histories," in *Tibetan Studies: Proceedings of the 6th Seminar of the International Association for Tibetan Studies*, vol. 1, ed. by Per Kværne (Oslo: The Institute for Comparative Research in Human Culture, 1994), 517–32. The geo-centrism is often connected with the ethno-centrism, though the latter has been discussed and critiqued more often in contemporary postcolonial scholarship. See Hwa Yol Jung, "Edouard Glissant's Aethetics of Relation as Diversity and Creolization," in *Postcolonialism and Political Theory*, ed. by Nalini Therese Persram (Lanham, MD: Lexington Books, 2007), 193–225.

As per the cross-cultural translating practice, in the early medieval China, some translators seemed to have modified some common expressions for fulfilling the expectations of Chinese readers, such as the order of the four directions, which was associated with the issue of ordering the four continents of the world. While exploring the organizing process of Buddhist translations, it should be noted that the monks and laypeople who were educated in traditional Chinese culture played important roles in transcribing, editing, and proofreading. However, foreign monks brought in and spoke out the original source versions of Buddhist literature. Some Chinese officials appointed by the emperor also followed imperial edicts to read and authorize the final versions. Therefore, it would not be surprising to see that the final versions of Buddhist translations used the order of the four continents that followed traditional Chinese literary conventions, which started from the east rather than from the south as the original version was produced in South Asia. The cultural geography in early medieval China impacted the vocabulary, sentence structure, and idioms of early Chinese translations of Buddhist literature.

In South Asian Buddhist literature, the four continents in Buddhist literature often refer to the continent in the south *Jambudvipa*, the continent in the west *Aparagodaniya*, the continent in the north *Uttarakuru*, and the continent in the east *Furvavideha*. In early medieval Chinese Buddhist literature, such as the *Biographies of Eminent Monks* (*Gaoseng zhuan* 高僧傳), both the land in the east (*dongtu* 東土) and the land in the middle (*zhongtu* 中土) referred to the Middle Kingdom, or in particular, the Central Plain, which was the central part of the Chinese empire. In the seventh and eighth centuries, some Chinese writers began to write about the four continents in the Chinese order, which started from the continent in the east. For instance, in his commentary on the *Vimalakīrti-nirdeśa-sūtra*, Huiyuan 慧遠 (523–592), a monk from the Pure Shadow Temple (Jingying si 淨影寺), said that the four continents under heaven started from the east and this continent was called *Furvavideha*,[15] which was a typical Sino-centric view. Later on, this Sino-centric order became even more common. In his *Record of the Western Regions of the Great Tang* (*Da Tang Xiyu ji* 大唐西域記), Xuanzang 玄奘 also introduced the idea that in Saha world (Skt. *sahāloka*) or the world of the Earth, there were four continents, namely *videha* (*Furvavideha*) in the east, Jambudvipa in the south, *Godani* (*Aparagodaniya*) in the west, and *Kuru* (*Uttarakuru*) in the north. However, Xuanzang was a Chinese monk who spent a significant amount of time studying in South Asia. He seems to be ambivalent while discussing the four continents, which means sometimes

15 *T.* 38, no. 1776: 514b.

he still holds South Asia in a higher place. When he commented on the features of these four continents, he again came back to the order starting from the south.[16] Xuanzang's understanding, perception, and conceptualization of the continent of Asia characterized the geography, climate, customs, norms, national characters, and economic modes of four peoples living in the land of Jambudvipa, which is Asia, as we can see. The landscape Xuanzang laid out seems to regard Mount Sumeru as the center of the world, and the order of four lands begins with South Asia, which refers to the land of the lord of elephants.[17] The order of arranging these four lands in Xuanzang's narrative might suggest that he still held South Asia as the birthplace of Buddhism prior to any other continent.

Although Xuanzang's writing followed the order of four lands starting from the south, many other monks in the high medieval period (tenth century and later) adopted the order of four continents starting from the east. For instance, in his commentary on the *Avataṃsaka-sūtra*, Chengguan 澄觀 states that the eastern continent looks like a half-moon. The southern continent looks like a carriage. The western continent looks like a full moon, and the northern continent looks like a square base.[18] In his *Outline of the Tian Tai Fourfold Teachings* (Tiantai sijiaoyi 天臺四教儀), a Korean monk named Chegwan 諦觀 (tenth century) also used the same order of the four continents starting from the east while discussing the life spans of people in the human realm. In the eastern land within the human realm, people could live to be 150 years old. In the southern land, people could live up to 100 years old. In the western land, people could live up to 500 years old. And in the northern land, people could live up to 1,000 years old. In these four continents, residents could enjoy happiness and endure suffering. While suffering, people should practice five ethics and five precepts. The five ethics

---

16 Samuel Beal trans., *Xi you ji: Buddhist Records of the Western World*, vol. 1 (London: Kegan Paul, Trench, Trübner & Co., 1906), 13–15.

17 On Xuanzang's positions about the Sino-India relations, see He Fangyao 何方耀, "Han Tang Zhongguo fomen de biandi yishi yu fanyu xuexi rechao 漢唐中國佛門的邊地意識與梵語學習熱潮," *Jiuzhou xuelin* 九州學林 3, no. 4 (2005): 137–40; Ge Zhaoguang 葛兆光, Zhaizi zhongguo: Chongjian youguan zhongguo de lishi lunshu 宅茲中國重建有關中國的歷史論述 (Beijing: Zhonghua shuju, 2011), 111–16, Chen Jinhua 陳金華, "Dongya fojiao zhong de biandi qingjie: lun shengdi ji zupu de jiangou 東亞佛教中的邊地情結: 論聖地及祖譜的建構 (The 'Borderland Complex' in East Asian Buddhism: a discussion of the development of sacred space and genealogies)," *Foxue yanjiu* 佛學研究 21 (2012): 22–41; Janine Nicol, "Outsiders: Medieval Chinese Buddhists and the 'Borderland Complex'—An Exploration of the Eight Difficulties," *The SOAS Journal of Postgraduate Research*, 6 (2014): 27–48.

18 *T.* 35, no. 1735: 583b.

were humaneness, righteousness, propriety, wisdom, and integrity, which of course, refer to the core values of Confucianism. The five precepts were no killing, no stealing, no sexual misconduct, no lying, and no intoxicants.[19] Interestingly, although the order of the four continents started from the east, based on this narrative, the people who lived in the northern land had the longest life span. The *Mūlasarvāstivāda Vinayavibhaṅga* translated by Yijing 義淨 also used the same order of the four continents within the human realm. It describes how to draw a painting of a wheel of life and death in five realms and the human realm; the four continents are marked in white color.[20]

As early as the sixth century, Chinese monks started to incorporate indigenous Chinese thoughts into their interpretations of the *MMS*. When Zhiyi 智顗 (538–597) wrote his *Great Calming and Contemplation* (Mohe zhiguan 摩訶止觀), he introduced the theory of Five phases (Wuxing 五行) while explaining the animal spirits in the *MMS*. He claimed that twelve beasts were cultivating their Buddhist compassion in the jeweled mountains and these beasts were lords of demonic spirits. For him, when a person was working on sitting in meditation, his or her mind might become deviant, which means that he or she fell into the seduction of the demonic animal spirits. When these demonic spirits attempted to seduce a Buddhist practitioner, they appeared in various body forms such as boys or girls, old adults, and birds or beasts. He continued to explain a strategy of distinguishing different animal spirits pursuant of fighting against their seductions. He suggested to follow the traditional Chinese idea of terrestrial branches for time divisions. One day could be divided into twelve units, and one unit has two hours. Within this two-hour unit, one animal demon would come to disturb the practitioner. For example, if it were the yin 寅 time unit, the tiger demon would come. If it were the chou 丑 time unit, the ox demon would arrive. Since there were more beasts and birds in nature, Zhiyi expanded the list of twelve animals in chapter 23 of the *MMS* to thirty-six beasts, which presented an entirely new cosmology in terms of time and space for animals as active spirits in a Buddhist context. According to Zhiyi, one two-hour time-unit can be divided into three subdivisions, so twelve time-units originally for twelve zodiac animals based on twelve terrestrial branches can accommodate thirty-six animals. Zhiyi's new interpretation is remarkable for its Chinese transformation of the twelve zodiac animals in the *MMS* in many ways. First, the tiger has replaced the lion in Zhiyi's list. Zhiyi's

---

19 *T.* 46, no. 1931: 776c.

20 T. 23, no. 1442: 811a. Stephen F. Teiser, *Reinventing the Wheel: Paintings of Rebirth in Medieval Buddhist Temples* (Seattle and London: University of Washington Press, 2007), 60.

list was also expanded from twelve animals to thirty-six animals, even one hundred and eight animals. Second, in Zhiyi's new explanation, the animals were no longer the transformations or reincarnations of Bodhisattvas for serving the needs of sentient beings; instead, they became demonic spirits. They could divert practitioners away from right consciousness and corrupt their minds. They could also speak about auspicious and bad omens. Third, Zhiyi now began to incorporate the Chinese indigenous thoughts of twelve terrestrial branches and Five phases (wuxing) into his theory. Twelve animals were regrouped based on five phases in his theory. These combinations and incorporation together seem to have transformed the Buddhist version of twelve zodiac animals as Bodhisattvas into a hybrid tradition mixed with Chinese theories of twelve terrestrial branches and five phases. Finally, Zhiyi also offered a strategy for eliminating the demon spirits. He said, if one knows the corresponding terrestrial branch for each animal demon, one could just call out the name of the animal, then the demonic beast spirit would disappear.[21] This operationalization of exorcism borrowed from early medieval Daoism.

From Twelve Animal Bodhisattvas to Twelve Demonic Spirits

The twelve zodiac animals appeared in various forms and images in medieval Chinese Buddhist literature. For instance, they appeared as animal guardians of twelve time units (shier shishou 十二時狩) in Wonhyo's 元曉 (617–686) Commentary on *The Treatise of Awakening of Faith in the Mahāyāna* (Dasheng qixin lun shu 大乘起信論疏) and Fazang's 法藏 commentary on *The Treatise of Awakening of Faith in the Mahāyāna* (Dasheng qixin lun yiji 大乘起信論義記). Yet, in the sixth century, they appeared as demonic spirits in Zhiyi and Zhanran's 湛然 commentaries on the passage from the *MMS*. The transformation from twelve guardians to twelve demonic spirits shows that the images of the twelve zodiac animals became negative and horrific with the appearance of new commentaries in the medieval period, which might indicate that some Chinese Buddhist commentators attempted to transform the original Buddhist tradition from South Asia into Chinese tradition in East Asia. This section will trace this transformation in the multireligious, intellectual, and cultural context of medieval China. I suggest that Buddhism, Daoism, and other Chinese elements interacted with each other and collectively contributed to this transformation.

In the *MMS* context, twelve calendrical animals appeared as the transformation bodies of virtuous Bodhisattvas. These Bodhisattvas could

21 *T.* 46, no. 1911: 115a–b.

appear as various body forms of all sentient beings in six realms such as gods (*devas*), demigods, human beings, nāgas, beasts, birds, and other animals, for the sake of enlightening other sentient beings. They traveled around the human realm or Jambudvipa, teaching, and transforming sentient beings. In other words, to teach and transform sentient beings, Bodhisattvas could appear in any form of a sentient being from one of six realms in Buddhist cosmology. This idea seems to be developed in the early Mahāyāna tradition, which means a Bodhisattva could come to this world in a transformed body for the purpose of helping sentient beings to eliminate karma and collect good deeds and merits. However, medieval Chinese commentaries by Zhiyi, Zhanran, Wonhyo, and Fazang, replaced the images of these Bodhisattvas with those of demon spirits, which seems to manifest the influence of the popular indigenous Chinese ideas of animal spirits in the early medieval period.

Different monks used different terms for the twelve zodiac animals in their commentaries on the *MMS* in medieval China. Zhiyi began to name these twelve animals as the enchanting calendrical demons (shimeigui 時魅鬼) or just calendrical enchantments (shimei 時魅), which means that, when these animals showed up, they would cast spells to deviate the mind of a Buddhist practitioner.[22] As Zhiyi illustrated in his *Essential Methods of Practicing and Cultivating the Calming and Contemplating by Sitting Meditation* (*Xiuxi zhiguan zuochan fa yao* 修習止觀坐禪法要), "Twelve calendrical animals could transform and change into various forms and materials, and some of them could appear in the forms of young ladies and old people, and even the horrible bodies and so forth, without a single form, deluding and tempting practitioners. These various spirits of enchantment (jingmei 精魅) who attempted to delude the practitioners would show up when their respective schedules came up. One should be capable of identifying them."[23] In Zhiyi's description, the order of the twelve calendrical animal spirits started from the tiger. They appeared to corrupt the minds of practitioners. As a solution for dealing with these animal spirits, Zhiyi said that the practitioners should identify these animals and loudly call out the names of these animal spirits; then the disturbance of these animal spirits could be eliminated.

22 While describing the blueprint of the Jetavana Monastery, in his notable *Illustrated Scripture of Jetavana Monastery* (Zhongtianzhu sheweiguo Qiyuansi tujing) Daoxuan mentioned that there was a ward within this monastery in charge of counting daily time for informing residents of the Buddhist schedule, which corresponded to the mode of the appearance and disappearance of twelve calendrical animals; see T. 45, no. 1899, 884c. For an English translation, see Antonino Forte, *Mingtang and Buddhist Utopias in the History of the Astronomical Clock: The Tower, Statue, and Armillary Sphere Constructed by Empress Wu* (Roma: Istituto Italiano per il Medio ed Estremo Oriente, and Paris: École française d'Extrême-Orient, 1988), 47–48.

23 *T.* 46, no. 1915: 470b.

Zhanran, another monk from the Tiantai tradition, followed Zhiyi and used the same term. Zhanran first pointed out that the so-called twelve calendrical animals in the *MMS* were the same as the twelve zodiac animals in five phases.[24] Apparently, Zhanran attempted to interpret the twelve animals in the *MMS* in terms of Chinese ideas of the five-phase theory. Since these twelve calendrical gods looked like animals, they were called calendrical animals.[25] Zhanran also expanded the number of calendrical animals to one hundred and eight to fit a grand theory, including the ideas of five phases, five cardinal directions, and three stages of each of four seasons.[26] He was perhaps the first Buddhist monk who connected twelve calendrical animals in Buddhist text *MMS* with the twelve zodiac animals in the indigenous Chinese cosmological tradition. On the one hand, Zhanran accepted twelve animals as calendrical animals in early Buddhism; on the other hand, he also introduced Chinese ideas of twelve zodiac animals and five phases in constructing a new concept of twelve animals in the *MMS*. In his commentary, he said that these animals appeared in accordance with

---

24 *T.* 46, no. 1912: 407b–c. For a study on Zhanran, see Jinhua Chen, "One Name, Three Monks: Two Northern Chan Masters Emerge from the Shadow of Their Contemporary, the Tiantai Master Zhanran (711–782)," *Journal of International Association for Buddhist Studies* 22, no. 1 (1999): 1–91; Chi Limei (Chi Reibai) 池麗梅, *Tōdai tendai bukkyō fukkō undō kenkyū josetsu: Keikei tannen to sono shikan fukōden guketsu* 唐代天台佛教復興運動研究序說：荊渓湛然とその止観輔行伝弘決 (Tokyo: Daizō Shuppan, 2008).

25 It should be noted that twelve calendrical gods also appeared as hourly gods in Manichaeism. In a Manichaen fragment found from Dunhuang now preserved at the National Library of China, it said that the twelve hourly gods could appear as either twelve bright kings or twelve female deities. See Édouard Chavannes and Paul Pelliot, "Un traité manichéen retrouvé en Chine I," *Journal Asiatique* 10, no. 18 (1911): 499–617; Lin Wushu 林悟殊, *Monijiao jiqi dongjian* 摩尼教及其東漸 (Manichaeism and Its Spread to the East) (Beijing: Zhonghua shuju, 1987), 217–29; The list of Manichean twelve hourly gods also appeared in Parthian, Sogdian, and Turkic manuscripts from Turfan, see Mary Boyce, *A Reader in Manichean Middle Persian and Parthian* (Leiden: Brill, 1975), 132; Hans-Joachim Klimkeit translated and presented, *Gnosis on the Silk Road: Gnostic Texts from Central Asia* (San Francisco: HarperSanFrancisco, 1993), 77–81.

26 In his *Xuanyuan Benji* 軒轅本紀 (Chronicles of the Yellow Emperor), Wang Guan 王瓘 noted how the calendar in the ancient times was made. The ancient sages combined the five phases, terrestrial branches, twelve two-hour time-units, and twelve zodiac animals for marking days, months, and years. The Yellow Emperor observed the Animal Tamer (Fuxi) drawing eight diagrams and counted three eight-diagrams for marking twenty-four solar terms for matching with four seasons. See Yunji qiqian 雲笈七簽, compiled by Zhang Junfang 張君房, vol. 100, Zhengtong daozang 正統道藏 (Daoist Canon compiled in the Zhengtong Reign Period), vol. 37 (Taipei: Xinwenfeng chuban gongsi, 1985), 100–10.

when they were scheduled to do so. Each hour one animal would show up. When the enchantment of an animal came up, or an animal demon came up, upon calling its name it would disappear. So Zhanran said that the animal demons were frightened when their names were verbalized by people.

Furthermore, the practitioners should not only know their names but also recognize their forms to prevent them from showing up to disturb people and commit malign acts. As I will show below, this idea and method of dispensing animal demons seems to be familiar in Chinese tradition in the early medieval period, emanating from the Chinese indigenous and Daoist traditions.

So why did Zhiyi and his later generations transform these twelve calendrical animals as the reincarnations of Bodhisattvas into demonic animal spirits? Or why did these Chinese Buddhist writers choose demonic spirits to replace the reincarnations of the Bodhisattvas as the new images of twelve zodiac animals? The root of the idea of demonic spirits was not from the Buddhist tradition, but Chinese indigenous culture. In Chinese Buddhist literature, the term "enchanted spirit" first appeared in a text entitled *Treatise on Śākya Mahāyāna* (*Shi moheyan lun* 釋摩訶衍論), which was attributed to Nagarjuna and was translated into the Chinese language by Vṛddhimata (Fatimoduo 伐帝摩多) in 401. This text states that there are four obstacles against mind cultivation that could appear in the form of human beings. These four obstacles were demons (Skt. *mara*, Ch. *mo* 魔),[27] non-Buddhist teachings (Skt. *anya-tīrtha*, Ch. *waidao* 外道), ghosts (*gui* 鬼), and gods (*shen* 神).[28] In the Buddhist context, demons refer to the demonic Mara and his associates, and the anya-tīrtha refers to non-Buddhist forces.

27 Alex Wayman, "Studies in Yama and Māra," *Indo-Iranian Journal* 3 (1959): 44–73; Travor Oswald Ling, *Buddhism and the Mythology of Evil: A Study in Theravada Buddhism* (London: Allen & Unwin, 1962); James W. Boyd, *Satan and Mara: Christian and Buddhist Symbols of Evil* (Leiden: E. J. Brill, 1975); Robert Warren Clark, "Mara: Psychopathology and Evil in the Buddhism of India and Tibet," PhD dissertation, University of Virginia, 1994; Bulcsu Siklos, "The Evolution of the Budhist Yama," in *Buddhist Forum*, vol. IV, ed. by Tadeusz Skorupski (London: School of Oriental and African Studies, University of London, 1996), 165–89.

28 In medieval Europe, there were three similar categories that often disturbed the human realm and society: demons, ghosts, and monsters. See Robert Bartlett, *The Natural and the Supernatural in the Middle Ages*, The Wiles Lectures given at the Queen's University of Belfast, 2006 (Cambridge: Cambridge University Press, 2008); David D. Gilmore, *Monsters: Evil Beings, Mythical Beasts, and All Manner of Imaginary Terrors* (Philadelphia: University of Pennsylvania Press, 2009); Timothy S. Jones and David A. Sprunger eds., *Marvels, Monsters, and Miracles: Studies in the Medieval and Early Modern Imaginations* (Kalamazoo: Medieval Institute Publications, 2002); Bettina Bildhauer, *The Monstrous Middle Ages* (Toronto: University of Toronto Press, 2004); Jean-Claude Schmitt, *Ghosts*

When the Buddha was meditating under the Bodhi tree, he was challenged, seduced, enchanted, and tempted by the Mara and various transformations and avatars of evil forces. Eventually, the Buddha defeated these evil forces and achieved enlightenment. However, these evil forces continued to bother other Buddhist practitioners. The Buddhist concept of animals as demonic spirits is illuminated in the *Treatise on Śākya Mahāyāna*. According to this text, ghosts refer to the so-called ten big ghosts and their numerous attendants, and spirits refer to fifteen big spirits and their numerous attendants. The tenth type of ghost is the ghost of animals, such as the scorpion, tiger, and lion. The fifteenth type of god is called the god of spirits of enchantment (jingmei shen 精魅神).[29] The gods of spirits of witchcraft appeared as animals as their avatars based on the daily schedule of these animals.

The Buddhist concepts of "mo 魔" and "waidao 外道" cannot be found in Chinese culture before Buddhism introduced them into China. Nevertheless, similar concepts of ghosts and gods can also be found in the indigenous Chinese tradition. As some contemporary scholars suggested, ghosts and demons in traditional Chinese society are the supernatural equivalent of bandits, beggars, and other dangerous strangers. They were considered to be marginal, dangerous, and capable of causing harm and disruption if ignored. The malevolent ghosts are turned from the deceased yet unrest souls of those people who did not fulfill the ideal life, especially those soldiers who died young on the battlefield, the poor people who died of hunger, or the unmarried maidens who have not yet fulfilled their social obligations of reproduction and were not fully accepted as normal family and social members in the traditional Chinese value system.[30] While they were not accepted in the society, they

---

in the Middle Ages: The Living and the Dead in Medieval Society (Chicago: University of Chicago Press, 1998); David Williams, *Deformed Discourse: The Function of the Monster in Mediaeval Thought and Literature* (Montreal: McGill-Queen's University Press, 1999); Jeffrey Jerome Cohen, *Of Giants: Sex, Monsters, and The Middle Ages* (Minneapolis: University of Minnesota Press, 1999).

29 For the most comprehensive study on the spirits of enchantment, see Lin Fushi 林富士, "Renjian zhi mei: Han Tang zhiji de jingmei gushi xilun 人間之魅：漢唐之際的精魅故事析論 (The Enchantment within the Human Realm: A Analytical Discussion on the Tales about the Spirits of Enchantment between the Han and Tang periods)," *Zhongyanyuan shiyusuo jikan* 中研院史語所集刊 78, no. 1 (2007): 107–82.

30 Arthur P. Wolf, "Gods, Ghosts, and Ancestors," in *Religion and Ritual in Chinese Society*, ed. by Arthur P. Wolf (Stanford: Stanford University Press, 1974), 131–82; Emily M. Ahern, *The Cult of the Dead in a Chinese Village* (Stanford: Stanford University Press, 1973); Maurice Freedman, "Ancestor Worship: Two Facets of the Chinese Case," in *Social Organization: Essays Presented to Raymond Firth*, ed. by Maurice Freedman (Chicago: Aldine Publishing, 1967), 85–103; Lin Liming 林禮明, Guiyu shijie 鬼魊世界 (The World of Ghosts and Demons) (Xiamen: Xiamen daxue chubanshe, 1993);

kept coming back to disturb the community of the living. Once they received sacrificial offerings, they would be in rest. This concept apparently received attention of medieval Chinese Buddhists, which was the reason why medieval Chinese Buddhists invented new discourses to reconcile the conflicts between Buddhist values and traditional Chinese family values. For example, the ghost festival was invented for feeding the female hungry ghosts. The male sons could save their mothers from the realm of hungry ghosts by devoting themselves to the Buddhist monastic community, becoming monks and serving the Buddhist community.[31] Poo Moo-chou noted that the concept of the ghost was sophisticated in the Warring States period, because the ghosts in ancient China might also come from animals, plants, and even woods, rocks, fire, and wind, etc., besides the unrest deceased humans might become ghosts.[32] Nevertheless, after Buddhism introduced new concepts of sentient beings in the realms of hungry ghosts and animals, there would be encounters between traditional Chinese concept of animal ghost and the Buddhist one (Figure 5.2).

The demonic spirits in Zhiyi's writing seem to suggest that the Buddhist idea of demonic spirits was mixed with the Chinese concept of ghosts. Why did Zhiyi and Zhanran choose the spirits of enchantment to interpret the animal avatars in the *MMS*? For me, they both attempted to make Buddhist teachings and ideas accessible and understandable to Chinese readers who were more or less familiar with the Chinese concepts of demonic animal spirits. In medieval Chinese literature, the animals as demonic spirits who could seduce and corrupt human minds seemed to have been a shared idea in both the Buddhist textual community and non-Buddhist textual community. Therefore, Zhiyi introduced this idea in his interpretation of the twelve zodiac animals in the *MMS*. Zhiyi might also utilize this idea to compete with the Daoist tradition since the latter often advocated that Daoist priests could exorcise animal demons by calling out their names in early medieval China.

Mu-chou Poo, *Ghosts and Religious Life in Early China* (Cambridge: Cambridge University Press, 2022); Mu-chou Poo, "The Completion of an Ideal World: The Human Ghost in Early Medieval China," *Asia Major* 3rd series, no. 10 (1997): 69–94; idem., "Ghost Literature: Exorcistic Ritual Texts or Daily Entertainment?" *Asia Major* 3rd series 13, no. 1 (2000): 43–64; Mu-chou Poo ed., *Rethinking Ghosts in World Religions* (Leiden: Brill, 2009).

31 Stephen F. Teiser, *The Ghost Festival in Medieval China* (Princeton: Princeton University Press, 1988).

32 Poo Mu-chou, "Zhongguo gudai gui lunshu de xingcheng 中國古代鬼論述的形成 (The Formation of the Concept of ghost in ancient China)," In *Guimeishenmo: Zhongguo tongsu wenhua cexie* 鬼魅神魔：中國通俗文化側寫 (Ghost, Enchantment, God, Demon: Perspectives on the Chinese Popular Culture) (Taipei: Maitian chubanshe, 2005), 30.

Figure 5.2: Dunhuang manuscript Pelliot chinois 4513, *Guanyin jing* (Scripture on Bodhisattva Guanyin) shows the animal-form demons. © The Bibliothèque nationale de France.

To trace this issue, I will briefly outline the idea of animism and its connection with the Buddho-Daoist ritual tradition in early medieval Chinese literature. Some early medieval Chinese literature, such as tales and histories, indicated the popularity of animism among monastic and lay communities, which believed that animal and plant spirits played active roles in daily life. In other words, there were differences between spirits and bodies. These spirits might appear in the body forms of humans, animals, and plants for disturbing the daily life of spiritual cultivators. This idea is different from what we saw earlier in the *MMS*. In the latter, the animals were Bodhisattvas but appeared as animal forms for teaching Buddhism and enlightening sentient beings. While these two ideas were encountered in early medieval China, some texts that appeared in this period seem to document this sort of encounter. For example, the *Scripture of Consecration* (*Guanding jing* 灌頂經) offers a list of numerous spirits and demons, such as spirits of trees and woods, spirits of worms, spirits of birds, spirits of beasts, and spirits of stoves and doors. It claims that if someone wore a talisman, the spirits of beasts (e.g., lions, tigers, wolves, and bears) would hide and harm nobody.[33] According to Michel Strickmann,

33 *T.* 21, no. 1331: 503a.

this scripture was a Buddhist apocryphal text that incorporated many indigenous Chinese thoughts into its content. Although some names of these spirits might originally come from the Buddhist tradition of South Asia, many names of spirits can be found in the Chinese indigenous tradition. Strickmann points out that the rise of this scripture in the six dynasties could be due to the spread of the thought of Final Dharma. The Final Dharma thought brought up the Buddhist sense of impermanence to the Buddhist community. The Buddhist community felt the threat of political and social disorder following the decline of Buddhist teaching. Therefore, evil spirits and demons appeared in the world to endanger the cultivation of Buddhist followers. *The Scripture of Consecration* taught both Daoist talismans and Buddhist spells (Dhāraṇī) for defeating these evil spirits.[34]

The tradition of writing about demonic spirits has a long history in ancient China. In ancient Chinese texts such as Zuo Qiuming's commentary on the *Annals of Spring and Autumn*, Du Yu's commentary on the *Record of the Grand Historian* (*Shiji*), and Yan Shigu's commentary on the *Book of the Han Dynasty* (*Hanshu*) the demonic spirits often refer to spirits who have the body forms of beasts for their claws and teeth that could harm or even deprive human lives. These beasts became so-called evil animals whose threatening images have appeared in numerous cultures across the world. For instance, in his commentary on the *Record of Grand Historian*, Du Yu noted that the demonic spirit had a human face, beast body, and four feet, which was produced by the exotic air of mountains and forests; and it often deluded humans. Typically, the evil spirits were portrayed as beasts with some human body parts, but with the nature of animality, not humanity, and their residence was associated with wilderness, such as mountains and forests.

It seems that different regions across China believed in different beast spirits. In the Southern Dynasties, evil spirits were said to appear as female bodies because they could enchant and delude other people. For example, in the *Book of the Chen Dynasty* (Chen shu 陳書), it is stated that Zhang Lihua, the concubine of the last lord of the Chen Dynasty, was talented with enchantment and could delude other people. She set up a cult in the court for controlling the lord.[35] In the meantime, a belief in the fox spirit was also developed, which was later mixed with the idea of

34 Michel Strickmann, "The *Consecration Sūtra*: A Buddhist Book of Spells," in: *Chinese Buddhist Apocrypha*, ed. by Robert E. Buswell Jr. and Kyoko Tokuno (Honolulu: University of Hawaii Press, 1990), 75–118; Michel Strimann, *Mantras et mandarins: le bouddhisme tantrique en Chine* (Paris: Gallimard, 1996), 127–63.

35 Yao Silian et al, *Chen shu*, juan 7 (Beijing: Zhonghua shuju, 1972), 131.

the female spirit.[36] Both Buddhism and Daoism in the Southern Dynasties accepted the idea that fox spirits could disturb their religious practitioners.

In the Northern Dynasties, cat and rodent spirits were more popular. In the Sui Dynasty, some women worshiped cat demons. Dugu Tuo (ca. sixth century), the stepbrother of Sui Emperor Wendi's (Yang Jian, 541–604, r. 581–604) wife, once tried to command a cat demon to enchant Wendi's wife, but his conspiracy was uncovered and thwarted. So, Wendi ordered that Dugu Tuo should be killed. Wendi's wife requested to save Tuo's life by fasting for three days. Wendi eventually followed his wife's request and enacted a less severe punishment.[37] The rodent demon was reported several times in the capital city when Wendi's son Yang Yong (562?–604) was the crowned prince. So, Wendi ordered one of his officials, Xiao Ji (525?–614?), to perform an exorcism for Yang Yong's palace to get rid of the rodent demons. Xiao Ji performed a Daoist ritual of exorcism by commanding gods to suppress these rodent demons.[38] It should be noted that many early medieval Chinese Daoist texts tell stories of exorcism. For instance, in its chapter on killing demons, a fifth-century Daoist text entitled the *Scripture of Divine Incantations of the Supreme Pervasive Abyss of the Most High* (Taishang dongyuan Shenzhou jing 太上洞淵神咒經, juan 14), notes that if the demon king transformed himself into the bodies of birds and brought up the pandemic in the human realm, Daoist priests could erect a Daoist altar for performing an exorcism. The Daoist masters of three caverns could create ritual space, turn scriptures, chant spells, set up a feast, offer sacrifices, and pray to the gods to end the pandemic and bestow blessings. Many other Daoist ritual manuals also offer ritual methods for exorcising beast demons. In the chapter on Petitioning to Officials of the *Book of Magic Writs of the Orthodox Oneness* (Zhengyi Fawen zhangguan Pin 正一法文章官品, DZ. 1218), the Daoist lord could command celestial generals and soldiers in charge of tiger and rodent demonic spirits.[39]

---

36 For a study on the fox spirit, see Xiaofei Kang, *The Cult of the Fox: Power, Gender, and Popular Religion in Late Imperial and Modern China* (New York: Columbia University Press, 2006).

37 Fang Xuanling et al. *Sui shu*, juan 36 (Beijing: Zhonghua shuju, 1973), 1108–9.

38 *Suishu*, juan 78, pp. 1775–76.

39 For a discussion on dealing with rat plagues in this text, see Franciscus Verellen, *Imperiled Destinies: The Daoist Quest for Deliverance in Medieval China* (Cambridge, MA: Harvard University Asia Center, 2019), 112. There are many stories about how Daoist Priests combated with the animal demons in Daoist literature in early medieval China, see Li Fengmao 李豐楙, "Liuchao jingguai dongwu yu Daojiao fashu sixiang 六朝精怪動物與道教法術思想 (The Demonic Animals and the Daoist Rituals and Thoughts in the Six Dynasties)," *Zhongguo gudian xiaoshuo zhuanji* 中國古典小說專輯 3 (1981): 1–36.

In the early medieval period, Buddhist ideas of reincarnation from humans to beasts were incorporated into some Daoist literature. Some Daoist texts said that deviant demons and beast spirits could corrupt men, and the beast spirits could seduce women. In the chapter on saving those in suffering of the *Scripture of the Most High from the Dongxuan Lingbao Canon Regarding Retribution and Karmic Causes* (Taishang dongxuan lingbao yebao yinyuan jing 太上洞玄靈寶業報因緣經, DZ. 336), a story tells that a tiger spirit deluded a woman who resided in the deep mountain, so she gave birth to four sons who had tiger bodies. Daoist priests helped her escape the delusion by offering spells, chanting scriptures, and bestowing precepts.[40] The *Scripture of the Most High from the Dongxuan Lingbao Canon Regarding Retribution and Karmic Causes* also noted that the humans who received bad retributions from their evil deeds could transform their bodies into many beasts, such as rabbits, deer, pythons, tigers, eagles, dogs, foxes, and snakes. If one received the body of a snake, he or she could always be thirsty and hungry and was hunted by everyone who saw it. This scripture claimed that if one loaned property or money to someone else but intimidated the debtors, he or she could be reborn in the body of domestic beasts, such as a cow, sheep, pig, or dog. Hence, reincarnation as beasts and being hunted and killed became common in Buddhism and Daoism in the early medieval period.[41]

Conclusions

This chapter started from a close reading of a message in the Chinese translation of the *MMS*. Twelve zodiac animals were portrayed as the reincarnated forms of Bodhisattvas who took turns to travel to different realms to teach those who were reborn in the life forms of animals and enlighten them. This message described a cosmological image of the Buddhist world centered on Mount Sumeru, with four continents and four seas surrounding this mountain. However, in this message, these four continents were arranged in a geographical order that started from the south international

40 For more discussions on tigers in medieval Chinese Daoism, see Huaiyu Chen, *In the Land of Tigers and Snakes*, chapter 4.

41 In medieval Europe, there are tales on the dog demons' transformation into martyrs after their conversion to Christianity; see Robert Bartlett, *The Natural and Supernatural in the Middle Ages* (Cambridge: Cambridge University Press, 2008), 94–95; John B. Friedman, *The Monstrous Races in Medieval Art and Thought* (Syracuse: Syracuse University Press, 2nd edition, 2000), 72–75; David Williams, *Deformed Discourse: The Function of the Monster in Medieval Thought and Literature*, 286–97; Joyce Tally Lionarons, "From Monster to Martyr: The Old English Legend of Saint Christopher," in *Marvels, Monsters, and Miracles: Studies in the Medieval and Early Modern Imaginations*, 167–82.

cardinal direction in the *MMS*. I suggest that this arrangement was due to the South Asia-centric worldview in early Buddhist cosmology. When the *MMS* was translated into the Chinese language, this South Asia-centered worldview was replaced with a Sino-centric worldview that considered the east as the first of four cardinal directions of four continents. The Chinese translation was processed and transmitted, which might indicate that in early medieval China, Buddhist writers and learners attempted to adapt Buddhist knowledge of geography and environment to meet the needs of Chinese Buddhist readers.

Furthermore, while commenting on the twelve zodiac animals in this message, many Chinese Buddhist writers, such as masters Zhiyi, Zhanran, and others, mobilized indigenous Chinese intellectual resources to interpret these animals. They attempted to establish corresponding relationships and connections between the twelve zodiac animals in the *MMS* and traditional Chinese theories of five phases and twelve terrestrial branches. In the meantime, they reinterpreted these twelve animals by introducing ideas of beast demons in the early medieval period across the boundaries of Buddhism, Daoism, and popular beliefs. Therefore, the original twelve animals in the *MMS* which appeared as compassionate Bodhisattvas later were transformed by medieval Chinese Buddhist writers into demonic beast spirits.

Appendix

Zhiyi's interpretation combining five directions, five phases, and thirty-six beasts[42]:

Directions	East	South	West	North	Center
Phases	Wood	Fire	Metal	Water	Earth
Beasts	Wild cat, leopard, tiger, fox, rabbit, racoon, dragon, alligator, fish	Cicada, carp, snake, deer, horse, water deer, sheep, goose, eagle	Ape, gibbon, monkey, bird, rooster, pheasant, dog, wolf, dhole	Hound, pig, mythical beast, cat, rodent, mythical cattle, crab, tortoise	Osprey, dhole, turtle
	寅 yin 卯 mou 辰 chen	己 ji 午 wu 未 wei	申 shen 酉 you 戌 xu	亥 hai 子 zi 丑 chou	

42 In common with other ancient ethno-biographical classification systems, some animals in the list are not identified as real nonhuman animals.

Chapter 6

THE WERE-TIGERS IN MEDIEVAL CHINA AND ITS ASIAN CONTEXT

Introduction

In his book *Anthropology and Religion*, Robert L. Winzeler offers a list of roles of animals in religious belief and practice, including animals as a god, deities, or cult creatures, animals as totems, animals as were-animals, animals as omen creatures, animals as tricksters, and animals as objects of taboo. While making examples of the were-animals, he particularly mentioned the werewolves in Europe and were-tigers in Indonesia and Malaysia.[1] Besides South and Southeast Asia, East Asia also had folklores on the were-tigers. After all, the were-tiger culture is deeply rooted in Asia's natural environment and ecology. Tigers were a significant apex predator across nearly half the Asian continent, which could be found historically in most regions of Central, South, Southeast, and East Asia. Unfortunately, the tiger became distinct from Central Asia in the tenth century and disappeared in Java and Bali in the twentieth century. Nowadays, only the Bengal tiger, Amor tiger, South China tiger, and Sumatra tiger still struggle to survive as endangered species in some nature reserves and zoos across South, Southeast, and East Asia. While interacting with humans, tigers not only appear as real nonhuman animals in the historical experience of human life

---

1 Robert L. Winzeler, *Anthropology and Religion: What We Know, Think, and Question*, 2nd edition (Lanham: Altamira Press, 2012), 81. Actually, in India some tribes such as Konda Reddis also developed folklores on were-tigers, see L. A. Cammiade, "Man-Eaters and Were-Tigers," *Man* 31 (1931): 217–20; in Naga Hills, the borderland of India with Burma, local people do not make distinction between tigers and leopards, so they have folklores on both leopard-men and tiger-men. They use one word "angshu" to refer to both tigers and leopards; see J. H. Hutton, "Leopard-Men in the Naga Hills," *The Journal of the Royal Anthropological Institute of Great Britain and Ireland* 50 (1920): 41–51. For the study on the were-tigers in Orissa, see Francesco Brighenti, "Kradi Mliva: The Phenomenon of Tiger-Transformation in the Traditional Lore of the Kondh Tribes of Orissa," *Lokaratna* vol. 4 (2011): 11–25.

but also play crucial roles in the spiritual and emotional experience of human life and culture. Contemporary scholars have studied tigers from agriculture, ecology, environmental history, and animal history perspectives.[2]

The study of the were-tigers in Asia can be traced back to the rise of European colonialism in the nineteenth century.[3] When Europeans began their adventures in the Asian continent, they gradually exposed themselves to the information about the tigers and were-tigers in Asia. They then developed preliminary studies on the were-tigers in the early twentieth century, undoubtedly inspired by their knowledge of the werewolves in Europe. In 1901, Jan J. J. de Groot first examined the were-tigers in Chinese history and literature in the ancient and medieval times in a chapter of volume four of his monumental book *The Religious System of China.* In traditional Chinese texts, he mainly introduced numerous stories about the transformations from tigers to humans.[4] Then nearly one century later, Charles E. Hammond analyzes the symbolic use of the tiger and were-tigers in the ancient lore of China.[5] Similarly, some other scholars in East Asia also studied various aspects of the were-tigers from the perspective of literary studies. In the past two decades, many scholars who focus on South and Southeast Asia made more progress in studying tigers in Asian religions, which significantly enhances our understanding of the significance of tigers in Asian culture. While inspired by these studies, I will look further at issues about the were-tigers in ancient and medieval China from a comparative perspective. On the one hand, some

---

2 See Robert B. Marks, *Tigers, Rice, Silk, and Silt: Environment and Economy in Late Imperial South China* (Cambridge: Cambridge University Press, 1998); Huang Zhifan 黃志繁, "Shanshou zhijun, huhuan yu daode jiaohua: cezhong yu Ming Qing nanfang shanqu 山獸之君、虎患與道德教化：側重於明清南方地區 (The Lord of Mountain Beasts, Tiger Disaster, and Moral Transformation: With Special Reference to Southern Regions in the Ming and Qing Dynasties)," *Zhongguo shehui lishi pinglun* 中國社會歷史評論 (Chinese Social Historical Review), vol. 7 (2006): 1–16; Chen Huaiyu, *Dongwu yu zhonggu zhengzhi zongjiao zhixu* 動物與中古政治宗教秩序 (Animals and Political and Religious Order in Medieval China) (Shanghai: Shanghai guji chubanshe, 2012); Hou Yongjian 侯甬堅, Cao Zhihong 曹志紅, Zhang Jie 張潔, and Li Ji 李冀, *Zhongguo huanjingshi yanjiu* 中國環境史研究 (Studies on Chinese Environmental History), vol. 3: Lishi dongwu yanjiu zhuanji 歷史動物研究專輯 (Special Volume on Historical Animals) (Beijing: Zhongguo huanjing kexue chubanshe, 2014).

3 For the introduction, See Patrick Newman, *Tracking the Weretiger: Supernatural Man-Eaters of India, China and Southeast Asia* (Jefferson, NC and London: McFarland & Co., 2012), 3–8.

4 J. J. de Groot, *The Religious System of China*, vol. 4 (1901), 162–81.

5 Charles E. Hammond, "Excursions in Tiger Lore," *Asia Major* 4.1 (1991): 87–100; and idem., "Sacred Metamorphosis: The Weretiger and the Shaman," *Acta Orientalia Hungaricae* 46, no. 2–3 (1992–1993): 235–55.

unique characteristics of the were-tiger narratives in medieval China were rooted in Chinese cosmological thoughts. On the other hand, the stories of the were-tiger in medieval China shared some similarities with those in other parts of Asia, though they might come from different historical periods.

This study intends to revisit the concerns surrounding were-tigers in China and contextualize them from a trans-Asian viewpoint in light of the "animal turn" and "materialistic turn" in modern humanities scholarship. First, this study will explore the confrontation between the transformation from tigers to humans in traditional indigenous Chinese thought and the reincarnation thought of Buddhist animals imported from South Asia. Second, this study is also concerned with the connections and similarities among the were-tiger narratives between China and other parts of Asia. In the recent decade, the new development of animal studies in the contemporary humanities reminded us of the significance of animals, whose agency kept shaping and reshaping our history and life. At the same time, they cohabit with humans on our planet.[6] The "materialistic turn" in animal studies, on the other hand, pays more attention to the bodies and effects of animals that impact human history and life, especially religious life.[7] While discussing the were-tigers, it is essential to understand the bodily transformation between tigers and humans as the change of life forms, which also brought up the emotional, mental, cognitive, and spiritual changes from tigers to humans. In this sense, it is worth looking into the discourses on life forms and related physiological and emotional issues in the indigenous Chinese thought. With the introduction and spread of Buddhism into China, some new discourses on the life forms and animals in Buddhist thought also spread into China. Then the Buddhist thought and the indigenous thought began to reconcile and adapt within the Buddhist community to fulfill the Chinese Buddhists' spiritual needs. Eventually, new discourses on the were-tigers appeared in medieval China.

In this chapter, I will first offer a reflexive retrospect on the conventional scholarship and point out the remaining issues. My study will shift the focus to intellectual history in the first part. Then I move on to the issue of

6 Erica Fudge, "A Left-handed Blow: Writing the History of Animals," in *Representing Animals*, ed. by Nigel Rothfels (Bloomington: Indiana University Press, 2002), 3–18; Hilda Kean, "Animal-Human Histories," in *New Directions in Social and Cultural History*, eds. by Sasha Handley, Rohan McWilliam, and Lucy Noakes (London: Bloomsbury, 2018), 173–89; Anna Peterson, "Religious Studies and the Animal Turn," *History of Religions* 56, no. 2 (2016): 232–45; Aaron Gross, *The Question of the Animal and Religion: Theoretical Stakes, Practical Implications* (New York: Columbia University Press, 2014).

7 Donovan Schaefer, *Religious Affects: Animality, Evolution, and Power* (Durham: Duke University Press, 2015).

reconciling two opposing narratives about were-tigers in China and other Asian countries. The final section will compare the Chinese were-tiger case with the South and Southeast Asia report. This research aims to uncover the connections between Asia's environment, history, animals, and cultures.

Some Reflections on the Contemporary Scholarship

Contemporary scholarship in many languages has discussed the were-tiger narratives in China comprehensively. In English-language scholarship, two scholars should be highlighted for their contributions to this topic. The first one is de Groot, who wrote a chapter on "Tigroanthropy" while discussing the soul issue in Chinese philosophy and popular ideas in the fourth volume of his six-volume work, *The Religious System of China*. He noted that in the vast land of China, the story about were-tiger almost exited in every province, with only a few exceptions, such as Shandong. He pointed out that in Chinese history, the first were-tiger in the textual sources was Gongniu Ai 公牛哀, which appeared in Liu An's *The Huainanzi* 淮南子.[8] He commented that the man-tiger such as Gongniu Ai in early China was also like the case described by European physiologists, which connected lycanthropy with the virulent disease, delirium, and insanity. The man-tigers in early China were also depicted as the one who was bloodthirst, craving human flesh and committing numerous murders and cruelties. The man-tiger was created in his imagination by fever patients who felt his limbs had grown longer or shorter, transforming him into the most ferocious monster. He then enumerates several other man-tiger stories found in Chinese textual sources.[9]

De Groot attempted to find the shared features of the man-tigers in ancient China and werewolves in Europe. For example, he noticed that the method of treating a man-tiger was to call his name so he could return to the human form. In ancient China, many people believed that talismans and exorcism on the personal names could pacify the man-tigers. Similarly, in Europe, a werewolf could change back to the human form if his baptized name was called. Citing the source Treatise on Curiosities (Bowuzhi 博物誌) in the third century, he described that some people in a brave tribe could turn themselves into tigers and transform back into a human. However, he could not hide their tails once they turned back to human forms, and they had to burn their tails to become humans. Then he connected this story with the tiger cults in Java and Sumatra, where some tigers could adopt human forms but could not hide some of their bodily features of tigers, such as the upper lips.

8 John S. Major, Sarah A. Queen, Andrew Seth Meyer, trans, *The Huainanzi*, 87.
9 de Groot, *The Religious System of China*, vol. 4 (1901), 162–65.

De Groot indicated that these people were regarded as "barbarians" for their ability to switch between the tiger and human forms. De Groot also noted that some tribes in Southwest China, such as the Ba people worshipped the tiger as their ancestral tribal totem. De Groot noted that on Mount Lamongan in Java, there was a village of man-tigers. The villagers there could transform into the tiger form, so they usually refused the outsiders for night-lodging because they were worried that their secret of dangerous power as man-tigers could be leaked. He continued to note that the villagers could also transform into tigers in Sumatra, a large village on the Dempu, a volcano in Palembang. The villagers from another neighboring village could even transform into tigers and pigs and then transform back.

Interestingly, De Groot touched on how the transformation could be completed. He suggested that in both Asia and Europe, charms, spells, and other instruments of witchcraft were used. In Europe, there were also witchcraft formulas, certain unguents, wolfskins, and mysterious rings or girdles for becoming a werewolf. In ancient China, similarly, the tiger skin appeared for an older man to transform into a tiger capable of grasping away livestock back home, though later he died in the wilderness from the thunder.

Dr. Groot paid attention to the changing souls of humans and the metamorphosis of beasts in the Chinese tiger lore. As he noted, while Gongniu Ai 公牛哀 turned into a tiger, his soul also changed. In the Eastern Han Dynasty, Feng Shao 封邵 became a tiger and began to eat humans, which indicated the changing soul too, not just the body transformation. De Groot also argued that the Chinese gentry class accepted the popular lore of the man-tigers as some stories appeared in the state-sponsored official histories, such as the *Old History of the Tang Dynasty* (Jiu Tang shu 舊唐書). Using some cases that appeared in the *Tiger Florilegia* (Huhui 虎薈) by Chen Jiru 陳繼儒 (1558–1639), a literati in the Ming Dynasty, De Groot suggested that the Chinese legend of the man-tigers and the werewolves in Britain and France shared a similarity that the bodily transformation from a human to a beast might be the outcome of the evil punishment. For example, some places worshiped the local cult for the river gods in ancient China. If the travelers did not pay homage to the gods, they could be punished by the god for transforming them into tigers as punishments. Nevertheless, De Groot pointed out that sometimes the Chinese were-tigers were harmless to humans as long as they received domestic livestock as sacrificial offerings, similar to the kind of female werewolves in Europe.[10]

10 de Groot, *The Religious System of China*, vol. 4 (1901), 165–82.

Hammond focused on the roles of tigers in Chinese thought and literature. He noted that the man-tiger cult in traditional China might receive the influence of non-Han tribes. He emphasized the importance of Shamanism in molding the lore of the man-tigers because of the similarities between Chinese and Southeast Asian lore of the man-tigers. One of the most important contributions of Hammond's study is the introduction of the feminist perspective for analyzing the tales about the were-tigers in medieval Chinese literature. He addressed the issue of female were-tigers in the Tang dynasty. He attempted to connect the appearance of female were-tigers with the rise of Wu Zetian 武則天, the first and only Chinese female emperor because the tiger represented the ferocious power of women. He also analyzed the Shaman background of a Tang tale about a woman called Xiaozhu 小珠. Xiaozhu was fed with raw meat and transformed into a tigress. The local villagers killed her. However, she was a Shaman. So she took avenge by mobilizing the tiger terror against the local villagers who participated in killing her. Hammond discussed this story from the feminist perspective. For him, Xiaozhu was depicted as a woman who suffered repression by male power. While attempting to overthrow this male power, Xiaozhu took the form of a tigress. This story happened in the southern region, which was marginal and remote to the center of the imperial authority, the court in Chang'an, north China. Hammond argued that Xiaozhu's rebellion as the marginal power of the barbarians and women challenged the dominance of a patriarchal society. Another story in his analysis depicted similar feminist thought. He noticed that since the Song dynasty, many were-tigers were female. In the Record of the Listener (Yijianzhi 夷堅誌), he raised three examples about wives as were-tigers. In these tales, wives suffered the dominant Confucian patriarchal ideology. Their jealous power grew up and transformed them into tigresses. So they became devouring people and attacked their servants and the petty wives of their husbands. Aside from female were-tigers, Hammond pointed out that many medieval Chinese legends about were-tigers reflected how minorities in Chinese culture were perceived. He analyzed some tales from *In Search of the Supernatural* (Soushen Ji 搜神記) that many were-tigers appeared among the minority tribes such as the Chu and Miao people in the frontier regions of the Chinese empire. Again, he linked these tales with Shamanism in these marginal regions. Like de Groot, he also suggested that the tales of were-tigers in medieval China shared some similarities with those of werewolves in medieval Europe. However, he attempted to explore the Chinese ideas behind these stories, for which he argued that in China, the transformation might be due to either the tiger's corruption of the human or Heaven's power. In addition, he suggested that, unlike Europeans who viewed werewolves as demons, the Chinese held more sophisticated views on were-tigers. In general, tigers were often regarded as charming and even sacred beasts. In ancient China, the were-tiger was regarded as a natural transformation. He indicated that even in medieval Chinese Buddhism, the appearance of were-tigers was not a direct consequence of Buddhist incarnation but a direct transformation. He claimed that the transformation of the were-tiger represented a shift from humanity to animalistic. Finally, Hammond tried to distinguish between the were-tigers as shamans and other types of shamans. For him, the were-tiger shamans could heal the ill people and save the dying victims. In the meantime, the were-tiger Shamans may also serve as the communication medium of Heaven.[11]

Contemporary scholarship in East Asia on the were-tigers primarily focused on the literary analysis of the tales in the medieval period.[12] Although there are numerous articles, I will highlight several most remarkable studies in the following. Japanese scholar Tominaga Kazuto 富永一登 focused on the tales of transforming into tigers in the Six Dynasties and the Tang period. He categorized four types of modes in the were-tiger transformation narratives. First, the cruelty of local officials led to the fear of local people toward the violent governance, so they developed the legend of local officials' transforming into tigers.

Second, the local people irritated the gods of the local temple cult, so the gods transformed into tigers to punish the violators. Third, some people were infected by fearful illness and turned themselves into tigers.

11 Charles E. Hammond, "The Demonization of the Other: Women and Minorities as Weretigers," *Journal of Chinese Religions* 23, no. 1 (1995): 59–80; idem., "Sacred Metamorphosis: The Weretiger and the Shaman," *Acta Orientalia Hungaricae* 46, No. 2–3 (1992–1993): 235–55.

12 Dong Naibin 董乃斌, *Zhongguo gudian xiaoshuo de wenti duli* 中國古典小說的文體獨立 (The Classical Novels as Independence Genre in China) (Beijing: Zhongguo shehui kexue chubanshe, 1994), 226; Cheng Guofu 程國賦, *Tang Wudai xiaoshuo de wenhua chanshi* 唐五代小說的文化闡釋 (The Cultural Interpretation on the Novels in the Tang and Five Dynasties) (Beijing: Renmin wenxue chubanshe, 2002), 150; Wang Fenling 汪玢玲, *Zhongguo huwenhua* 中國虎文化 (Chinese Tiger Culture) (Beijing: Zhonghua shuju, 2007), 371; Wang Li 王立. "Liaozhai zhiyi Huagao huahu fuchou gushi de Zhong Yin wenxue yuanyuan 聊齋誌異向杲化虎復仇故事的中印文學淵源 (The Sino-Indian Literary Sources for the Tale of the Tiger Transformation of Xiang Gao for Taking Revenge in the Strange Tales from a Chinese Studio)." *Huanan shifan daxue xuebao* 華南師範大學學報, no. 1 (2008): 59–66; For a recent study, See Long Sheng 龍聖 and Li Xiangzhen 李向振, "Binghuan: Bianpo gushi de shehui yinyu 病患: 變婆故事的社會隱喻 (The Illed Patients: The Social Metaphor of the Tales on the Human Transformations)," *Minzu wenxue yanjiu* 民族文學研究 (Ethnical Literature Studies) 37: 3 (2019): 5–15. They all did not mention Japanese scholarship.

Finally, some "barbarians" living on civilization's outskirts turned into tigers.[13] Some of these points shared similar observations with de Groot and Hammond, as I mentioned earlier. Shimamori Tetsuo 島森哲男 noted that there were only three modes of the tales about were-tigers in medieval China: humans transformed into tigers, tigers transformed into humans, and human-tiger marriage. He argued that these transformations mainly depended on wearing or taking off clothes, which indicated the difference between humanity with clothes and animalistic without clothes. He also paid a little attention to the Daoist idea that the gods could command animals as their messengers and its impact on the tales of were-tigers.[14] Although many articles in Chinese touched on this theme, the most comprehensive study is a Master's degree thesis completed by Hong Ruiying 洪瑞英 in 1991. This thesis examines the thoughts, contents, and mental consciousness in the were-tiger tales. It first attempts to define the concept of the transformation from human form to tiger form. It limits the definition of the body transformation, especially the transformation with the tiger skin. This thesis points out that the early thought of the tiger transformation was connected with the totem of the cult, which means that the humans might embody some characteristics of the totem animals. However, with the development of the tales in the Six Dynasties, many tales on the were-tigers reflected more contemporary ideas and projections on society, human life, morality, and mentality. Hong also tried to connect the tales of the were-tigers with the widespread phenomena of tiger violence in medieval China. She categorized the following types of the tales on the were-tigers: officials transforming into tigers, human-tiger marriage, transformation by divine power, transformation by karmic attributions,

---

13 Tominaga Kazuto 富永一登, "Jin toraten no keifu: Rikuchō katoraten kara Tōten shōsetsue 人虎傳の系譜ー六朝化虎譚から唐傳小說へ (The Geaology of the Weretiger Tales: From the Tales of The Tiger Transformation to the Stories in Tang Legends)," *Chūgoku chūsei bungaku kenkyū* 中國中世文學研究, 13 (1978): 1–17; Kakuitsu Shoji 荘司格一, "Jin toraden seiritsu no haikei chusei ni okeru tora no setsu 人虎伝」成立の背景—中世における虎の説話 (The Background of the Rise of the Weretigers: Tales on the Tiger in the Middle Ages)," *Yamagata daigaku kiyo* 山形大学紀要 9: 4 (1981): 417–43.

14 Shimamori Tetsuo 島森哲男, "Chugoku no hito tora henshintan: 'nugu/Kiru' no minwagaku 中国の人虎変身譚ー「脱ぐ / 着る」の民話学," *Miyagi Kyōiku Daigaku kiyō* 宮城教育大学紀要 no. 53 (2018): 403–26; Tominaga's articles did not appear in his reference list. Tokura Hidemi 戶倉英美, "Bianshen gushi de bianqian: you liuchao zhiguai xiaoshuo dao Liaozhai zhiyi 變身故事的變遷：由六朝志怪小說到聊齋誌異 (The Evolution of the Tales of Body Transformation: From the Anomaly Tale in the Six Dynasties to the Strange Tales in a Chinese Studio)," in *Guoji liaozhai lunwen ji* 國際聊齋論文集 (Collected Papers from the International Studies on Strange Tales from a Chinese Studio), eds. by Gu Meigao 辜美高 and Wang Zhizhong 王枝忠 (Beijing: Beijing shifan xueyuan chubanshe, 1992), 161–98.

transformation by witchcraft or exorcism, transformation by changing ether and essence, tiger grandmother, and tiger deviant demon, etc.[15]

These studies have indeed dealt with some most important sources and offered a comprehensive analysis of critical issues. However, some remaining issues are worth further discussion. For example, it might be interesting to look into the changing ideas on the were-tigers before and after the spread into China. Buddhist thought on animals has deeply rooted in its cosmology. Buddhism often understands and explains animals as sentient beings in its theory of the six realms of life and death. It viewed animals as inferior life forms to humans. While Buddhism was spread to China, its ideas of animals were confronted with indigenous thought in China. New ideas from reconciliation were developed by Chinese Buddhists who attempted to target their Chinese audience and make their interpretations acceptable to this audience. This study will examine these issues in light of the animal and materialist turn of contemporary scholarship. Furthermore, many contemporary studies on the were-tigers in China also touched on the comparative perspective with the werewolves in Europe, but the discussion is limited. Given that in the past three decades, many scholars who worked on tigers in South and Southeast Asia have made tremendous progress, it might be worth benefiting these studies by revisiting some connections between the were-tiger tales in China and those in other parts of Asia.

The Buddhist Encounter with Indigenous Chinese Thought

Since many aspects of were-tigers in ancient and medieval China have been discussed by contemporary scholars, my research in this section will focus on the encounters between Buddhist and indigenous Chinese thought on were-tigers and the transformation of tigers, particularly the encounter between Buddhist reincarnation and indigenous Chinese Yin-Yang theory. Although the first were-tiger narrative appeared in the *Huainanzi* compiled by Liu An, which was dated in the Han Dynasty and did not mention Buddhism at all, it happened almost coincidently at the same time Buddhism was introduced to China. While in the Northern and Southern Dynasties, the Buddhist ideas could be found in many Chinese narratives on the were-tigers. Given that the spread of Buddhism had a tremendous impact on Chinese politics, economy, culture, and the way of life in the early medieval period, it was essential to explore how genuine Buddhist ideas were accepted

15 Hong Ruiying 洪瑞英, *Zhongguo renhu bianxing gushi yanjiu* 中國人虎變形故事研究 (A Study on the Tales on the Transformations between Humans and Tigers in China) (Xinbei: Huamulan wenhua chubanshe, 2011).

in China. Some Buddhist ideas, such as Buddhas, Bodhisattvas, four noble truths, and reincarnation, were entirely new to the Chinese audience. Other ideas, such as the realm of beasts, were not entirely new. Although six realms and the cycle of life and death were introduced to China and the realm of gods, demigods, and hell was entirely new, the idea of beasts in Buddhism might find some counterparts in China. After all, some similar animal species, such as tigers, monkeys, horses, and cattle, also lived in China. As a result, the idea of human–beast physical transformation took a different route in China than that of nirvana and reincarnation. This path might be full of Buddhist reconciliation and adaptation in the medieval Chinese context.[16]

There were two intellectual traditions about the tiger transformations in medieval China. one was the indigenous tradition of transforming animals into spirits or demons in pre-Buddhist China.[17] The other was the Buddhist tradition of reincarnation in light of sentient beings in six realms. After the spread of Buddhism into China, with Buddhism took root in Chinese society, the Chinese Buddhist community gradually developed new narratives on animal transformation by incorporating indigenous Chinese traditional animal transformation to reinterpret Buddhist narratives of reincarnation. These two traditions also reflect the difference between the Buddhist understanding of sentient beings in various life forms and the Chinese understanding of life forms. Buddhists and Chinese traditions understand that humans and animals have different material or physiological forms of bodily materiality. In the meantime, both Buddhism and Chinese traditions paid attention to the spiritual dimension with the changing bodily materiality during the transformation. However, Buddhism often emphasized the impermanence of five aggregates as the main constituent elements of life.

---

16 For a discussion, see Chen Huaiyu, *Dongwu yu zhonggu zhengzhi zongjiao zhixu* 動物與中古政治宗教秩序 (Animals and Political and Religious Order in Medieval China) (Shanghai: Shanghai guji chubanshe, 2012), chapter 4 on the monk–tiger relationships.

17 Contemporary scholarship has done a lot studies on the spirits and demons; see Li Fengmao 李豐楙. "Liuchao jingguai dongwu yu Daojiao fashu sixiang 六朝精怪動物與道教法術思想 (The Demonic Animals and the Daoist Rituals and Thoughts in the Six Dynasties)," *Zhongguo gudian xiaoshuo zhuanji* 中國古典小說專輯 vol. 3 (1981): 1–36; Sawada Mizuno 澤田瑞穗, *Chūgoku dōbutsutan* 中國動物譚 (Chinese Animal Tales), Tokyo: Kōbundō, 1978; Liu Zhongyu 劉仲宇, *Zhongguo jingguai wenhua* 中國精怪文化 (Chinese Spirit and Anomaly Culture) (Shanghai: Shanghai renmin chubanshe, 1997); Tu Cheng-sheng 杜正勝, "Gudai wuguai zhi yanjiu: yizhong xintaishi he wenhuashi de tansuo 古代物怪之研究：一種心態史和文化史的探索 (A Study on Object Spirits in Ancient China: An Investigation from the Perspective of the History of Mentality and Cultural History)." (Three parts). *Dalu zazhi* 大陸雜誌 (Mainland Journal) 104, no. 1 (2001): 1–14; 104, no. 2 (2001): 1–15; 104, no. 3 (2001): 1–10.

Some stories about the tiger transformation were collected into the Buddhist encyclopedia compiled by medieval Chinese Buddhist masters. However, not all of these stories mentioned Buddhist influence. For example, in the Buddhist encyclopedia, The *Pearl Forest of Dharma Grove* (Fayuan zhulin 法苑珠林), compiled by Daoshi 道世 in the seventh century, many stories on the tiger transformation seemed not to mention Buddhist thought at all. It cited *Garden of Anomalies* (Yiyuan 異苑), a story in which a two-foot tiger was found in Nanyang 南陽 in 285. The story merely stated that the tiger was Yin 陰 in essence and resided in the Yang 陽 position, which was a beast of metal (jinshou 金獸). Nanyang was the name of fire. So the metal essence entered the fire and lost its original shape, indicating the court disaster's deviance.[18]

Here the story used the indigenous Chinese thought of the Yin-Yang and Five-phase theories. The tiger was viewed as the "Yin" and the "metal" beast. The *Pearl Forest of Dharma Grove* also cited a story about Yi Ba 易拔, a clerk from Yuzhang 豫章 Prefecture (modern Nanchang, Jiangxi Province) in the Jin Dynasty who turned into a giant tiger with three feet. Again, it did not mention any Buddhist element. In the meantime, this encyclopedia cited a story from the *Continued Record in Search of the Supernatural* (Xu soushen Ji) that the "barbarians" residing in the mountains in Xunyang County 尋陽縣 (modern Huangmei, Hubei Province) could use witchcraft to transform people into tigers. The practitioner of this style of witchcraft drew a giant tiger on a piece of paper and placed a talisman alongside the animal. Then he wore a three-foot turban and performed the witchcraft with a red rooster and a pot of alcoholic liquor. Although these cases appeared in the Buddhist encyclopedia, they did not explicitly mention the role of Buddhism.

As a Buddhist encyclopedia, the *Pearl Forest of Dharma Grove* was compiled to target the Chinese audience to convince and convert them. Thus, many stories in this encyclopedia attempted to teach Buddhist ideas and values. One story was connected with the Buddhist ideas of reincarnation and retribution. In this story, a monk living in the Sanguan pavilion outside the north gate of Wuchang City was killed by a tiger. He was still a layperson and made a living by planting vegetables in a garden.

A pig came to disturb the garden. Then he killed the pig. However, the pig reincarnated as a tiger but appeared as a man wearing black robes to devour the monk.[19] In this story, the deceased pig reincarnated to be a tiger and could also appear as a man in black and spoke to the monk in human language. Therefore, it seems to be a sophisticated story about the transformations

---

18 *Fayuan zhulin* 法苑珠林, T. 53, no. 2122, 531c.

19 Jiexian 戒顯, *Xianguo suilu* 現果隨錄, *Zokuzokyo* 續藏經, X 1642, 88, 35b.

between beasts and humans. In other words, it did fall into the category of Buddhist six realms in which sentient beings from the animal realm could be reborn in the human realm. Another story was about Wu Daozong's 吳道宗 mother. She transformed into a tigress in 408 because of her bad karma from her previous life. She then brought the disaster to her village and killed several villagers. Later a hunter wounded her, and she eventually died at home.[20] This story indicated that a person who had terrible karma in his or her previous life might result from the reincarnation of being reborn as a tiger or tigress, which was a typical Buddhist idea. These two stories reflected the acceptance of the Buddhist idea of reincarnation in early medieval Chinese tales. As sentient beings, humans might reincarnate as tigers or other beasts for their evil deeds from a previous life, in light of the Buddhist concept of life, death, and rebirth. This concept is different from the indigenous Chinese thought on the tiger as the metal beast with the Yin essence. The latter was based on the traditional Chinese Yin-Yang and Five-phase theories.

From the cases above, we can see the differences between traditional Chinese tales on the tiger transformation and the stories with Buddhist influence. However, some tales show that medieval Chinese Buddhism attempted to reinterpret the animal transformation narratives with the Buddhist thought of reincarnation. Some Chinese Buddhists aimed to proselytize the idea of cause and effect, so they often claimed that the transformation from humans to beasts verified the Buddhist efficacy of changing life forms. While raising some examples from the pre-Buddhist Chinese tradition, medieval Chinese Buddhists incorporated these examples into the Buddhist tradition for the Chinese audience. For example, in the Sui Dynasty, Li Shiqian 李士謙 used to be a filial son in a traditional Confucian family. Later, he converted to Buddhism. So when he talked with others about the Buddhist thought of reincarnation, he raised several examples of the animal transformation in ancient Chinese history, such as some historical and legendary figures' transformations into beasts, but attributed all of them to the Buddhist verification the reincarnation thoughts.[21] He stated that these transformations from humans to beasts were due to the deeds of the mind. He must have been familiar with some of these tales from the pre-Buddhist era before he became a Buddhist, but he reinterpreted them with Buddhist thought after converting to Buddhism.

20 *Fayuan zhulin* 法苑珠林, T. 53, no. 2122, 532b.

21 *Fozu lidai tongzai* 佛祖歷代通載, vol. 10, T. 49, no. 2036, 559b; *Longxing biannian tonglun* 隆興編年通論, *Zokuzokyo* 續藏經. 75, no. 1512, 153b, a list shows that these figures transformed into birds, dragon, tiger, pig, dog, turtle, cattle, fish, etc.

Since the Buddhists in medieval China reshaped the animal transformation narratives by introducing the Buddhist thought of reincarnation, this effort of the Buddhist reshaping impacted the texts in the late imperial periods. For example, in his work *Tongyi* 通翼, a Buddhist layperson Zeng Duanfu 曾端甫 explained the Buddhist thought of reincarnation by combining both indigenous Chinese classics and Buddhist scriptures. One chapter of his book dealt with protecting the lives of sentient beings. He answered a guest about reincarnation from the cases of goats eating people and people eating goats. He claimed that the discourse of reincarnation and the discussion on cause and effect should be viewed from two dimensions. Those cases seemed ridiculous in the dimension of the matter or phenomena (shi 事), while in the dimension of reasoning or rationale (li 理), they were logical. He suggested that humans and animals both had physiological bodies. Sometimes humans might transform into a horse and a python, which were often the transformation of the afterlife. However, Feng Shao's transformation into a tiger and Mingzhen's transformation into a snake were the transformation in this life.[22] His main concern was protecting sentient beings' lives, so his discussion was not limited to the tiger transformation. However, he attempted to interpret these animal transformations from the traditional Chinese philosophical notions such as phenomena and rationale. In the late Ming Dynasty, a Buddhist monk, Ruchun 如純, attempted to synthesize Buddhist and Confucian ideas to defend Buddhism against Catholicism. He noted that even before Buddhism entered China, it was not rare that humans could transform into animals. These transformations were not from the Buddhist thought of reincarnation but the thought of ancient Chinese sages on the changing essences and traveling souls. He first pointed out that the bodily transformations from humans to beasts did not begin with the introduction of Buddhism into China. For him, ancient Chinese sages said that the essence and pneuma constituted the physical matters, and the traveling souls were constantly changing. While a sentient being was alive, the sentient being had a mind; once the sentient being died, there was still a soul. So the mind and soul were not two different things. He stated that the transformation in Sage's theory and the reincarnation theory in Buddhism was the same.[23] Thus, he seemed to have borrowed the Neo-Confucian concept of the one principle in all things.[24]

22 Peng Jiqing, *Jushi zhuan* 居士傳, juan 47, *Zokuzokyo*. 88, no. 1646, 272b.

23 *Shengchao poxie ji* 聖朝破邪集 (1639), juan 8, *Dazangjing bubian* 大藏經補編 (Supplementary Volumes to the Buddhist Canon), 28, no. 155, 343a.

24 For the discussion on the one-principle thought in Neo-Confucianism, see Yong Huang, "Cheng Yi's Moral Philosophy," in John Makeham ed., *Dao Companion to Neo-Confucian Philosophy* (Dordrecht: Springer, 2010), 71–73.

The discussion above focused on the connections between Buddhist thought of reincarnation and the tiger transformation narratives. When it comes to the Buddhist community's response to the issue of indigenous Chinese philosophy, it's not unexpected that the Buddhist community recognized that the traditional Yin-Yang and Five-phase theories played vital roles in interpreting animal transformations as early as the Six Dynasties. The traditional Chinese cosmology centered on these theories already impacted the Buddhist community for many narratives in the Buddhist community, including the changing images of zodiac animals in the Buddhist texts.[25] In the following discussion, I will first offer a survey on the intellectual background of the traditional narratives of the tiger transformation and then examine how the Buddhist community theorized the tiger transformation narratives by incorporating the Yin-Yang and Five-phase theories into the Buddhist interpretation.

Before the spread of Buddhism into China, it seemed that in the indigenous Chinese tradition, there was a discourse on the animal transformation, which focused on the changing essence, pneuma, and the traveling souls.[26] From some narratives I mentioned above, the "mere-tiger" was said to come from the metal essence (Jinjing 金精), according to the traditional theory of changing essence and pneuma. The metal essence was one of the five pneumas (metal, wood, water, fire, and earth) that constituted the physical realm. The nature of the metal essence was said to be violent, which corresponded to the violent nature of animalistic. Therefore, the body from the metal essence could transform into the tiger form. Here it shows that traditional Chinese thought suggested that the materiality of the animal was connected with the essence and pneuma in the physical realm.

In the *Huainanzi*, chapter three on "Celestial Patterns" explained the issues of cosmology and positional astronomy. It claimed that the cosmos' original qi (pneuma) could be clear and bright or heavy and turbid. The pure bright pneuma spread out to form Heaven, while the heavy and turbid pneuma congealed to form the earth. Then "the conjoined essences of Heaven and Earth produced Yin and Yang. The supersession essences of Yin and Yang caused the four seasons. The scattered essences of the four seasons created a myriad of things. The hot *qi* of accumulated Yang produced fire; the essence

25 Huaiyu Chen, "The Changing Images of Zodiac Animals in Medieval Chinese Buddhist Literature," *Nesir Journal of Literary Studies* vol. 1 (2021), 37–54. See Chapter 5 in this book.

26 For a perspective on the humans as animals in pre-Buddhist China, see Michael Nylan. "Humans as Animals and Things in Pre-Buddhist China," *Religions* 10, no. 6 (2019): 360, p. 4; doi:10.3390/rel10060360.

of fiery *qi* became the sun. Then cold *qi* of accumulated Yin produced water; the essence of watery *qi* became the moon."[27] Then it explained how animals appeared. It said, "Hairy and feathered creatures make up the class of flying and walking things and are subject to Yang. Creatures with scales and shells make up the class of creeping and hiding things and are subject to Yin. The sun is the ruler of Yang. Therefore, in spring and summer, animals shed their fur; at the summer solstice, stags' antlers drop off."[28] According to this worldview, birds and beasts were Yang animals, whereas worms and fish were Yin animals. As the traditional Chinese cosmology explained, there were correlative relations among animals, cardinal directions, four seasons, five sounds, five colors, and five phases.[29]

Earlier in the *Master Lü's Spring and Autumn Annals* (Lüshi Chunqiu), a third-century encyclopedia, the chapter on "Resonating with the Identical" briefly noted the corrective relations between rulers and five *qi*. It said, "When it came to the time of Tang 湯, Heaven first caused metal blades to appear coming forth from Water. Tang proclaimed, 'The ethers of Metal are in ascendance.' Since the ethers of Metal were ascendant, he honored the color white and modeled his affairs on Metal."[30] This book also highlighted how metal was linked to political power. The vast historiographer, it was stated, would inform the Son of Heaven of "the Power that is thriving is Metal" once the Autumn was formed. Then the Son of Heaven began his purification. "On the Day beginning Establishing Autumn, the Son of Heaven personally leads the Three Dukes, the Nine Ministers, the feudal lords, and the grand officers in welcoming autumn at the western suburban altar. On returning, he rewards the commanding generals and military officers in ceremonies at court. The Son of Heaven then mandates that his generals and commanders should select men, sharpen weapons, choose and drill those of distinction and merit, and give their entire trust only to men of achievement, thereby correcting every immorality."[31] This information correlated the Autumn season with the power of metal and connected them with military power, violence, and weapons. These characteristics were the same as the nature of

27 *The Huainanzi*, by Liu An, translated by John Major et al., 115.

28 *The Huainanzi*, 116. For the study, see John S. Major, *Heaven and Earth in Early Han Thought: Chapters Three, Four, and Five of the Huainanzi* (Albany: State University of New York Press, 1993).

29 A. C. Graham made a chart to illustrate their correlative relations, see his *Yin-Yang and the Nature of Correlative Thinking* (Singapore: Institute of East Asian Philosophies, 1986).

30 *The Annals of Lü Buwei*, by Buwei Lü, Translated by John Knoblock and Jeffrey Riegel (Stanford: Stanford University Press, 2000), 283.

31 *The Annals of Lü Buwei*, 173.

the tiger in the eyes of ancient people, which reflects the direct observations of the ancient people and their historical and physical experience with the environment, animals, and society.

In the Han dynasty, the philosophy on the correlative relations among cosmic powers, humans, and animals became more comprehensive and systematic. The five stars corresponded with five phases. So the metal star (jinxing 金星) was connected with the white tiger, which was the ominous sign for the events in the human realm.[32] The white tiger was believed to be in charge of weapons and war, which was violent and ferocious.[33] Many texts in the Han Dynasty, such as *Garden of Eloquence* (Shuoyuan 說苑), the *Book of Han Dynasty* (Hanshu 漢書), and *Discussive Weighing* (Lun heng 論衡), all offered some discussions on the correlative relations between five stars and five phases. In general, the west, as the cardinal direction, corresponded with the element of metal, with the spirit of Taibai 太白, and the symbolic animal was the white tiger. In the Han dynasty, many stories in the official history indicated that there was the ether of metal in the Autumn season, which might be when the nomadic tribes attacked from the west and led the war disaster to the Han regime. Therefore, the ruler was admonished by his officials to ease the burden of his ruled subjects by reducing taxes and public duties. The ruler should also offer fire sacrifices in the western suburb of the capital city because the fire element could overcome the metal element according to the Five-phase theory. Many apocryphal texts and divination techniques in the Han dynasty accepted similar discourses and attempted to offer advice to the rulers. In sum, the corrective theory, combined with the Five-phase theory and the Yin-Yang theory in the Han dynasty, provided the historical and intellectual background for the rise of the tiger transformation narratives in early medieval China.

Since the idea that the ether of the metal essence caused the were-tiger transformation was widespread during the Han dynasty, it's not surprising that the Buddhist community used it to interpret Buddhist reincarnation while dealing with the were-tiger transformation, especially with the rise of anomaly tales during the Six Dynasties. Eventually, even the Buddhist eminent master paid attention to the idea of the metal essence and its connection with the were-tigers. In his encyclopedia, *The Pearl Forest of the Dharma Grove*, Daoshi

---

32 Ho Peng Yoke, *Chinese Mathematical Astrology: Reaching Out to the Stars* (London: Routledge, 2004), 49. On the stars and corresponding spirits, also see Michael Puett, *To Become a God: Cosmology, Sacrifice, and Self-Divinization in Early China* (Cambridge: Harvard University Asia Center, 2002), 109–17.

33 Ching-lang Hou, "The Chinese Belief in Baleful Stars," in *Facets of Taoism: Essays in Chinese Religion*, eds. by Holmes Welch and Anna Seidel (New Haven and London: Yale University Press, 1979), 209–19.

mainly listed a section on animal transformation. He cited Gan Bao *In Search of the Supernatural* to establish the correlative relations between five *qi* and five essences. Chapter 12 of the *In Search of the Supernatural* says, "There were five prime energies (*qi*) in Heaven – from these, all things were formed. The wood phase clear is benevolence; the fire phase clear is ritual behavior; the metal phase clear is righteousness; water clear is intelligence, and the clear earth phase is contemplation. When all five energies reach their apogee of clarity, qualifications for sage virtue are present." It also claimed that the metal phase being beclouded, we have cruelty and concluded that "More sages are engendered in the Central Lands because their harmonious forces intersect. The lands beyond reach engender many anomalies born of their exotic energies." Then it pointed out, "If a human begets a beast or a beast begets a human—this is energy in chaos. It is an energy transition when a man becomes a woman, or a woman becomes a man. Niu Ai 牛哀 of Lu fell ill and seven days later became a tiger. His whole body was transformed, tooth and claw grew huge, and when his older brother opened the door to enter, the younger sprang upon him and ate him. While he was human, he did not know he would be a tiger; while he was a tiger, he was unaware of once being a man."[34] However, after introducing the traditional Chinese thought on the were-tiger transformation, Daoshi turned to explain the Buddhist thought of karmic retribution. He argued that sentient beings' body shapes were formed by original consciousness and karmic deeds. Since the seed of the original consciousness reached maturity, its affinity would rely on the external form. Then the changing external elements would lead to the changes in the sentience and non-sentience of a sentient being so that the physical shape would also be transformed. It seems that Daoshi introduced both the traditional Five-phase theory and the Buddhist thought.

The Were-Tigers in Trans-Asian Cultures

Some tales of the were-tigers in medieval China seem to have entered the neighboring countries with the spread of Buddhist tradition. Some were-tiger tales in medieval China also appeared in the literary collections of Korea and Vietnam. In Japan, since the tiger was not a native animal, the were-tiger tales were not as popular as in Korea. Instead, there were many tales of the werewolves in Japan.[35] While in South and Southeast Asia,

---

34 Gan Bao, *In Search of the Supernatural: The Written Record*, translated by Kenneth J. DeWorskin and J. I. Crump, Jr. (Stanford: Stanford University Press, 1996), 142–43.

35 Hishikawa Akiko 菱川晶子, *Ōkami no minzokugaku: jinjū kōshōshi no kenkyū* 狼の民俗学: 人獣交渉史の研究 (Tokyo: Tōkyōdaigaku shuppankai, 2018).

contemporary scholarship has revealed that there are various traditions found in many regions. For example, in Orissa, India, some people could become the were-tigers because the earth goddess granted them supernatural power. This transformation reflected the willingness of the local deity. When the earth goddess felt she did not receive sufficient offerings from local villagers, she mobilized her supernatural power to transform some villagers into tigers to punish local people. So these were-tigers served as the spirit mediums between the earth goddess and local villagers.[36] In the Dempo region of Sumatra, Indonesia, local Java villagers believed that the lake area on the top of a local mountain was the territory of the Lord of Souls, Raje Nyawe, and other supernatural beings as the ancestors of the tribe. These ancestors might not pass away, but they could appear as were-tigers and communicate with local villagers.[37] In the Perumahan Panji Laras Indah region of Java, the local villagers also believed that female deities, were-tigers, and elders protected villages. The were-tigers mainly served to prevent the real tigers from accessing the villagers. The locals in this area often believed that the soul could transform from a human to a tiger body.[38] Although there are many traditions of were-tigers in other parts of Asia, some of them are even similar to what we know about the were-tigers in the Chinese sources, as I will reveal in the following discussion.

Interestingly, a tale about Shentu Cheng's 申屠澄 wife as a were-tiger in the *Extensive Records of the Taiping Xinguo Period* (Taiping Guangji 太平廣記, Juan 429) was rewritten by the writer in Silla and appeared with a new title "Kim Hyeon's being Moved by a Tiger maiden (Kim Hyeon Gamho 金現感虎)" in the Memorabilia of the Three Kingdoms (Samguk yusa 三國遺事, Juan 5). This tale combined Buddhist Jataka narratives, local Korean culture, and the were-tiger tale. In this modified version of Silla, it said that during the reign period of King Wonseong of Silla (r. 785–798), a young man Kim Hyon followed the contemporary Buddhist atmosphere for making merits by circumambulating the stupa of the Heunglyunsa 興輪寺. He came upon

---

36 Francesco Brighenti, "Kradi Mliva: The Phenomenon of Tiger-Transformation in the Traditional Lore of the Kondh Tribes of Orissa," 11–25.

37 Bart Barendrgt, "A Supernatural Topography of the Southern Sumatran Highlands," *Crossroads: An Interdisciplinary Journal of Southeast Asian Studies* 18, no. 1 (2006): 113–25, esp. 119–20.

38 Robert Wessing, "The Last Tiger in East Java: Symbolic Continuity in Ecological Change," *Asian Folklore Studies* 54 (1995): 191–218, esp. 198; idem., "Homo Narrans in East Java: Regional Myths and Local Concerns," *Asian Folklore Studies* 65, no. 1 (2006): 45–68, esp. 55; idem., "'Bangatowa', 'Patogu', and 'Gaddhungan': Perceptions of the Tiger among the Madurese," *Journal of Southeast Asian Studies* 25, no. 2 (1994): 368–80, esp. 377–79.

a maiden circumambulating the stupa late at night. Then fell in love with the shared values and practices.

The maiden then told him that she wanted to go home. Kim Hyeon went with her together to a shelter near the mountain. An old lady there told the maiden that she was not supposed to bring a stranger with her. So the maiden hid Kim Hyeon. A while later, three tigers roared to arrive in the shelter, and they told the old lady in human language that they were hungry and needed meat. The old lady and the maiden scolded the tigers. At that time, a voice from Heaven said that these tigers had already harmed many lives so that Heaven would kill one of them as punishment. Three tigers felt fearful of this. But the maiden told three tigers that she could replace them to receive the punishment because these three tigers were her older brothers. So three tigers just fled away. Then the maiden entered the shelter and told Kim Hyeon that she was not a human being but had a genuine love for Kim Hyeon. However, she would like to sacrifice herself to save her brothers. She claimed that Kim Hyeon would rather kill her than die at the hands of others because she would like to repay the love of Kim Hyeon. She made a plan and told Kim Hyeon that she would go to the market next and alert the people of her presence. While she also told Kim Hyeon to go after her to the forest. Kim Hyeon said he was not interested in sacrificing his lover for a reward from the king. However, she insisted that her short-lived life was her fate, and she could help her tiger family. She asked Kim Hyeon to build a temple to transfer the merits. The next day, she appeared as a tigress in the city. King Wonseong ordered to kill her. Kim Hyeon volunteered to hunt her down and followed her into the forest, where the tigress transformed herself into a maiden again. She told Kim Hyeon to forget the previous love and then killed herself with Kim Hyeon's knife. Later, Kim Hyeon passed the civil examination and served the government. He built a temple named the Tiger's Vow Temple (Howonsa 虎願寺).[39] This tale seems to remind us of a Jataka tale about a nine-color deer which was the previous life of the Buddha. With great compassion, the deer sacrificed himself to save the hunter whom

39 Many Chinese and Korean scholars have noted the modification of this story in Korea. See Jin Chengyu 金程宇, *Yuwai hanji congkao* 域外漢籍叢考 (Studies on the Chinese Texts in Foreign Regions (Beijing: Zhonghua shuju, 2007), 69–70; Chen Yiyuan 陳益源 and Xu Jincheng 徐錦成, "Lun yuwai hanwen biji xiaozhong de hu 論域外漢文筆記小說中的虎 (A Study on the Tiger in the Chinese Tales in Foreign Regions)," in *2007 Dongya hanwenxue yu minus wenhua guoji xueshu yantaohui lunwenji 2007* 東亞漢文學與民俗文化國際學術研討會論文集 (Collected Papers from the International Conference on East Asian Chinese Texts and Folklore Culture), eds. by Wang Sanqing and Chen Yiyuan (Taipei: Lexue shuju, 2007), 425–41.

the king would kill if the hunter could not hunt down the deer. In the meantime, Kim Hyeon's tale combined two elements: the mutual transformation between a tiger and a human being and the trans-species love story between a tiger and a human being. Although this story modeled the tale about Shentu Cheng's wife in the Chinese tradition, it does not mention anything about the traditional Chinese Yin-Yang and Five-phase theories. It merely delivered the Buddhist message that the tiger's killing of human lives would result in karmic retribution and punishment. As a result, the were-tiger tales in Silla were selectively rewritten and adapted to suit the Korean audience.

While in Vietnam, the were-tiger tale was connected with the supernatural power of the eminent monks for healing illness. A fourteenth-century collection of tales titled *Strange Tales Plucked from the Dust of the Lingnan region* (Lĩnh Nam chích quái 嶺南摭怪) collected a tale about how an eminent monk Nguyễn Minh Không 阮明空 (1065–1141) helped heal the were-tiger illness of Vietnamese emperor Lý Thần Tông 李神宗 (1116–1138). Once upon a time, Lý Thần Tông became ill and behaved like a ferocious tiger. Minh Không was invited to the court to help treat the emperor. He set up a huge pot, boiled oil, and then used the oil for bathing the emperor's body. The were-tiger illness was healed. Minh Không was granted the title of "the state master (Quốc Sư 國師)."[40] Later on, he was honored as a Bodhisattva (Không Lộ Vương Bồ Tát 空路王菩薩) by his Buddhist followers. This tale illustrated that the were-tiger transformed from a human was an illness that could be pacified by an eminent monk who could mobilize his supernatural power.

Another fascinating feature of the were-tiger transition is the link between Southwest China's were-tiger history and surrounding South Asian regions. Many scholars have noted that some tribes in southwestern regions, such as Yi (Nuosu), Qiang, and Tibetan tribes, developed some tiger cults. The tiger skin was often used to decorate the state and military power authorities and appeared in the funeral rites for honoring the deceased tribal members.[41] In the early medieval Chinese collection *In Search of the Supernatural*, a tale said that in the Chu 貙 tribal region near Changsha (modern Hunan Province), there were both tigers and were-tigers in the mountains. The local government used traps for hunting down tigers to save trouble from tigers'

40 Dai Kelai 戴可來, Yang Baoyun 楊保筠, Wu Shiying 武世營 eds., *Lĩnh Nam chích quái deng shiliao sanzhong* 嶺南摭怪等史料三種 (Three Historical Sources including Lĩnh Nam chích quái) (Zhengzhou: Zhongzhou guji chubanshe, 1991), 40–41.

41 Zhang Zehong 張澤洪, "Zhongguo xinan shaoshu minzu zongjiao zhong de huchongbai yanjiu 中國西南少數民族宗教中的虎崇拜研究 (A Study on the Tiger Cult among the Minority Religions in Southwest China," *Zhongnan minzu daxue xuebao*, no. 6 (2007): 38–43.

attacks. However, it often appeared that the were-tigers might fall into traps. One method for identifying these were-tigers was checking out their foot toes because "when you find a tiger with five toes, you have a Chu tiger."[42] The were-tigers had five toes, which indicated they were local tribal members who transformed into tigers rather than real nonhuman tigers in the natural environment. Although this story does not mention Buddhism at all, a Buddhist text titled *Samādhi-Sūtra has spoken by the Buddha on the Supreme Brightness greater than that of Sun* (Foshuo chaoriming sanmeijin 佛說超日明三昧經) translated by Nie Chengyuan, it mentioned that the Chu tribal man might transform into a tiger and could not be aware of being an huma being. While returning to harm the home, he could not recognize his relatives.[43] Therefore, again the were-tiger narrative was mixed with the Buddhist tradition in early medieval China.

This identification method by finding whether the were-tiger had five toes is remarkable. A similar method can be found in some other South and Southeast Asia regions. In this sense, this study can be expanded in a trans-Asian context. However, we might see the materials were from different historical periods, and they were of no direct concrete connections. In his study of the were-tiger tradition in India, Desmond Kharmawphlang pointed out that the tiger was viewed as the symbol of speed, strength, and brevity in Indonesia, Malaysia, Laos, Vietnam, Cambodia, and Burma. In many regions of Asia, some tribes believed that there were spiritual connections between wild animals and human beings and that the beasts also had souls.[44] Kharmawphlang noted that in the Khasis of Meghalaya in northeast India, local people believed the were-tiger or the tiger-man existed, which might be rooted in the widespread local animism. All sentient beings were generally viewed as beings with an eternal essence called "Ka rngiew," which shaped human behaviors, thoughts, motivations, dreams, and wishes. In the mountains, some tribes believed that some were-tigers might bear the essence of the tiger's Ka rngiew, so they called these were-tigers "Khla Phuli." Interestingly, these Khla Phuli were also called "San saram," literarily "five claws," for they have five claws in their hands and feet. And these "San saram" made them different from the real nonhuman tigers in the wilderness. These local villagers acknowledged that real tigers often had four toes. Although the two tribes lived in different historical times, this

42 *In Search of the Supernatural*, 148.
43 T. 15, no. 638, 545a.
44 Desmond Kharmawphlang, "In Search of Tigermen: The Were-Tiger Tradition of the Khasis," *India International Centre Quarterly* 27, no. 4/ 28, no. 1 (Winter 2000/Spring 2001): 160–76.

method of detecting were-tigers is strikingly similar to that used by the Chu tribes in early medieval China. Kharmawphlang found that the religious practice of this region was centered on the female-dominated society. The residents in this region traced their female ancestors back to "Iawbei" and believed that there was a tutelary deity called "ryngkew" (tiger) in every village, rock, and river. The were-tiger tradition seems to be a clan tradition, only existing among female clan members.

A similar phenomenon in the were-tiger tradition can be found in the Naga Hills region near the border between India and Burma. Local tribes also traced their ancestor back to the tiger or leopard. They believed the first spirit, first tiger, and first man came from the same single mother. So the villagers might have lived with were-tigers and were-leopards in the same environment. The method of finding out who was the were-tiger or were-leopard was to check out the footprint. The were-tiger and were-leopard often left the five-toe footprints. The locals thought that when a human's spirit fell asleep at night, it could connect to the body of a tiger or a leopard, but that it would return to a human's body during the day.[45]

Conclusions

To summarize the preceding discussions, this chapter has attempted to advance the debates on were-tiger transformations by documenting the cosmological and philosophical foundations in traditional Chinese thought, and their intertwined relationships with Buddhist reincarnation and karmic attribution thought in the medieval period. Before Buddhism was spread to China, the indigenous Yin-Yang and Five-phase theories laid the cosmological and philosophical foundations for the transformations between animals and humans. A strong tradition of correlative thinking flourished in ancient China during the Han Dynasty. The material realm in this thinking system was formed by five *qi*, which also constituted the material and spiritual realms of animals and humans, combined with five essences. Due to its ferocious character, the west regarded the tiger as the animal of metal essence in terms of the cardinal direction. This concept was widely accepted in the ancient and early medieval periods and appeared in various texts. Since Buddhism arrived in China, it soon accepted this idea and adapted it for interpreting its doctrine of reincarnation to make its religious teachings accessible to the Chinese audience. Therefore, it appeared that medieval Chinese

45 J. H. Hutton, "Leopard-Men in the Naga Hills," *The Journal of the Royal Anthropological Institute of Great Britain and Ireland* 50 (1920): 41–51.

Buddhist society combined indigenous Chinese thoughts with Buddhist ideas to develop new narratives on the were-tigers. With the spread of Buddhism to some neighboring regions, the were-tiger narratives also entered Silla and Vietnam. In Silla, the Buddhists modified the were-tiger narrative by eradicating the Chinese element while emphasizing its Jataka origins. Yet, in Vietnam, the narrative was used to preach the supernatural power of the Buddhist healing

Because historically, tigers were widely active in South, Southeast, and East Asia, the were-tiger narratives widely appeared in vast areas of Asia. There were numerous manifestations of the were-tiger narratives in various literature and folklore materials across Asia. These manifestations are often highly local but occasionally share some standard features. For example, my discussion revealed that the local villagers in many parts of South, Southeast, and East Asia used either five toes or five-toe footprints to identify the were-tigers and create a boundary between them and the nonhuman tigers in nature. From a trans-Asian perspective, the were-tiger narratives might indicate that in many tribes of South, Southeast, and East Asia, there might be the totemism of the tiger or the tiger cult. They believed their ancestors might be tigers in the mountains, or they acknowledged that they shared the living realms with the tigers in the wilderness. In some regions, the villagers also believed that these were-tigers represented the power of their ancestors to serve gods to protect them.

Chapter 7

THE ANIMAL TURN IN ASIAN STUDIES AND THE ASIAN TURN IN THE ANIMAL STUDIES

Introduction

As I have briefly noted in the introduction, my discussions in this book might interest readers who work on Animal Studies and Asian Studies. Although the "animal turn" did not appear first in Asian Studies, it seems that this turn has arrived in the contemporary scholarship of Asian Studies. More books and articles on animals in Asia have become available gradually from time to time in the past decade. Beyond America and Europe, more publications in Asia also pay attention to animal studies. At least two journals in Taiwan and mainland China, such as *Cheng Kung Journal of Historical Studies* 成大歷史學報 and the *World History Review* 世界歷史評論, organized special issues on animal history. Other special issues on posthuman history also published several articles on animal history. In addition, a publisher in Shanghai is planning a book series on animal history for the first time in the Chinese-language scholarly community. In the meantime, it is equally astonishing to witness the growing interest in Asian Studies in the field of Animal Studies. For instance, the animals and religion group within the American Academy of Religion strongly promoted panels and papers on animals and Asian religions at almost every annual conference. My observations from these conferences suggest that numerous scholars in religious studies have realized that Asian traditions played vital roles in understanding the sophisticated relationships between animals and humans and their implications for shaping our history and future shared with animals. Various animal studies networks in America, Asia, and other parts of the world often included papers and presentations on animals in Asian religions and history or literature.

As a concluding section, this chapter offers my critical assessment of a batch of important publications that have greatly enriched contemporary scholarship in animal studies and Asian studies. These publications particularly illustrate that animal studies is a transdisciplinary trend in Asian Studies that has crossed the traditional boundaries among the studies of history, literature, and religions. I have selected six representative publications to evaluate how the authors combine both new approaches in animal studies and traditional methods of textual analysis and close reading to reveal multiple facets of the cultural, medical, symbolic, and spiritual roles and functions of animals in traditional and contemporary religious, literary, and historical cultures in Asia. Furthermore, this book has also covered the biological, textual, philosophical, spiritual engagements of humans in medieval China with various plants, which illustrates multispecies relations.[1] The following discussions also touched on both animals and plants. In the end, combining both these reviews and precedent chapters, I will briefly summarize three approaches for the prospective scholarship on animal studies.

Animal Studies in Asian Religion and Literature

Scholars in the Humanities have seen an "Animal Turn" in contemporary Western-language scholarship in the past two decades. This has been due to contemporary political, economic, and ethical concerns about climate change, the crisis facing the environment and species, and the impact of these issues on the sustainable development of our planet. Numerous conferences and programs have been developed across the traditional disciplinary boundaries of philosophy, literature, history, religious studies, and art history. In particular, there has been a boom in the study of the relationship between animals and religions.[2] Much of this attention has been on the Abrahamic religions. However, for a comprehensive understanding of the sophisticated relationships between animals and religions, more studies are needed of animals and religions in the so-called Global South. Reiko Ohnuma's new book, *Unfortunate Destiny*,[3] makes a crucial contribution to current discussion on animals in religions, particularly Buddhism.

With her exceptional analytical skills, Ohnuma demonstrates how a close reading of Buddhist narratives as a method of traditional textual study can

1 Readers may find more theoretical issues in the following two books: Donna J. Haraway, *When Species Meet* (Minneapolis and London: University of Minnesota Press, 2008), and Euardo Kohn, *How Forests Think: Toward an Anthropology beyond the Human* (Berkeley: University of California Press, 2013).

2 Anna Peterson, "Religious Studies and the Animal Turn," *History of Religions* 56, no. 2 (2016): 232–45.

3 *Unfortunate Destiny: Animals in the Indian Buddhist Imagination* (New York: Oxford University Press, 2017).

Buddhist society combined indigenous Chinese thoughts with Buddhist ideas to develop new narratives on the were-tigers. With the spread of Buddhism to some neighboring regions, the were-tiger narratives also entered Silla and Vietnam. In Silla, the Buddhists modified the were-tiger narrative by eradicating the Chinese element while emphasizing its Jataka origins. Yet, in Vietnam, the narrative was used to preach the supernatural power of the Buddhist healing

Because historically, tigers were widely active in South, Southeast, and East Asia, the were-tiger narratives widely appeared in vast areas of Asia. There were numerous manifestations of the were-tiger narratives in various literature and folklore materials across Asia. These manifestations are often highly local but occasionally share some standard features. For example, my discussion revealed that the local villagers in many parts of South, Southeast, and East Asia used either five toes or five-toe footprints to identify the were-tigers and create a boundary between them and the nonhuman tigers in nature. From a trans-Asian perspective, the were-tiger narratives might indicate that in many tribes of South, Southeast, and East Asia, there might be the totemism of the tiger or the tiger cult. They believed their ancestors might be tigers in the mountains, or they acknowledged that they shared the living realms with the tigers in the wilderness. In some regions, the villagers also believed that these were-tigers represented the power of their ancestors to serve gods to protect them.

reveal the complexity of animals' participation in human religious life. Ohnuma also successfully introduces some new issues from the perspectives of material culture and effect studies,[4] discussing the roles the description of colors and sounds of animals have in Buddhist narratives for invoking religious feelings and emotions. As primary sources, Ohnuma focuses on Buddhist narratives, especially the Jātaka stories in the Pāli canon, along with the Sanskrit materials whenever they are available. The book's three parts are divided into seven chapters, with an appendix listing animal rebirths. Each part has an introduction to its structure and theme.

Part I has two chapters. Drawing mostly upon various Buddhist sutras, Chapter 1 demonstrates that the Buddhist view of animals does not differ from that of the West. In the Buddhist doctrinal and cosmological framework, animals are ontologically inferior to humans in terms of their mental faculties, moral agency and spiritual potential, lacking language, reason, intelligence, wisdom, and insight. Chapter 2 analyzes stories about the spiritual life of several animals—including a bull, parrot, snake, and buffalo—in Buddhist texts and argues that some animals as Buddhist converts in Buddhist discourses are capable of catapulting up the karmic hierarchy to be reborn into the highest level of Buddhism's six realms of life and death; however, the animals' abilities of moral agency, self-cultivation and enlightenment can only be gained by a denial of their animality.

Part II turns to the Jātaka stories to discuss the animality of animals in the complexity of Buddhist discourses, while also attending to the rhetorical and didactic functions of animals as allegorical human beings in the early Buddhist *imaginaire*. In this section, Ohnuma is primarily concerned about how, in the stories, animals are given voices to speak about the suffering of animals and humans. Chapter 3 explores themes of the cruel and abusive treatment of humans toward animals, who cook, eat, hunt, sacrifice, punish, and execute them. While discussing the trial and punishment of animals, Ohnuma compares the Buddhist case with the execution of animals in the early modern West, and she reveals that, in Buddhist rhetoric, dogs were sometimes given powers of speech, rational argument, and legal reasoning to advocate their case in order to preserve Buddhist justice. She highlights a very important distinction between the individual and whole species of animals as revealed in Buddhist discourse. A small section in this chapter

4 For the materialist turn in religion, see Aaron S. Gross, *The Question of the Animal and Religion* (New York: Columbia University Press, 2015); and for the effect studies in religion, see Donovan Schaefer, *Religious Affects: Animality, Evolution, and Power* (Durham: Duke University Press, 2015).

is a reminder that one narrative could translate ordinary animal sound into human language to illuminate the Buddhist morality of defending animals. Chapter 4 examines the Bodhisattva as an animal to illustrate his virtues of wisdom, compassion, and generosity as a savior. However, more importantly, Ohnuman shifts the focus to the animal as the Bodhisattva as a departure from the human-centered perspective and argues that, in some stories, the animal was depicted in Buddhist narratives as a thinking, speaking, and rational being in order to condemn the wickedness of human beings.

The three chapters in Part III are case studies of the unique roles of animals as the animal doubles of the Buddha. For Ohnuma, these animals are named and individualized, and cannot therefore be viewed as generic representatives of the animal species. Furthermore, these animals played unique roles in the life story of the Buddha, which marked some turning points in his spiritual journey. Chapter 5 offers a detailed discussion on the role of the horse Kanthaka, based on many different versions of the Buddha's life stories. Ohnuma argues that Kanthaka was portrayed in order to play multiple roles in the Buddha's renunciation experience. In light of Kimberley Patton and Wendy Doniger's studies on sacrifice, Ohnuma demonstrates that Kanthaka not only served as the symbol of the Buddha's final link with this worldly life as he rode him for his great departure and left him for an ascetic life in the forest; he also served as a sacralized sacrificial animal upon the Buddha's going forth, as well as a scapegoat when the horse returned to city. Chapter 6 deals with the parallelism between the Buddha and the elephant Pārileyyaka and argues that the latter was portrayed as the mirror of the former, who appeared as consummate Renouncer and Teacher in contrast to the monastic assembly. However, Ohnuma suggests that the elephant was still submissive to humans in the Buddhist discourse. The Buddha still maintained his dominance over animals, the world of Nature, and even the cosmic world. In the last chapter, Ohnuma presents the story of the Buddha's taming of the maddened elephant Nālāgiri and argues that this animal served as a billboard for the Buddha's power, charisma, and masculinity. She also compares her study with that of Garry Marvin on the Spanish bullfight. Finally, she analyzes the complicated relationships among the lion, elephant, and jackal in the stories of the Buddha by suggesting that the lion is associated with the Buddha's preaching of the Dharma, the elephant is associated with the Buddha's awakening, and the jackal refers to miserable, lower sentient beings.

Unfortunate Destiny is admirably lucid in its writing style, and should become an ideal, and popular choice for teaching courses on the relationships between animals and religions.

***

Animal studies has been flourishing across the boundaries of many humanities disciplines in the past decade. Vertebrate animals, particularly those who closely interact with humans, attract more attention than other animals. Invertebrate animals gradually draw some attention in the recent decade.[5] Wilt L. Idema's new book *Insects in Chinese Literature* is the first comprehensive volume on the images of various insects in Chinese literature through the ages, with a focus on the premodern period.[6] Although the subject on animals in Chinese literature has appeared in both Asian and Western-language scholarship, very few books and articles have been devoted to insects, or mostly invertebrates. In this sense, Idema's new book occupies a unique position of adding a welcoming reference to our understanding of animals in Chinese literature and even Asian literature in general. Furthermore, for the first time, through lucid and concise translations, this book makes many obscure Chinese popular literary texts accessible to the English-speaking world.

Idema starts his book by noting modern writer Lu Xun's famous essay on the insects in the Hundred-Plant Garden at his parents' home. He traced it back to Lu's connection with a Dutch novel *De kleine Johannes* perhaps via its German version during Lu's studying medicine in Japan in the early twentieth century. He then categorizes Chinese literature before the Literary Revolution in 1917 and the May Fourth Movement in 1919 into three traditions based on the practical means and purposes in the daily life of readers: texts for study and recitation, texts for entertaining reading, and texts for performance and listening. The first category refers to the tradition of *Belles Lettres*, the second category refers mostly to classical and narrative literature, and the third one refers to popular literature. Idema's book divides the majority of texts he studied and translated into two big groups: *Belles lettres* texts (nine chapters for Part I) and popular literature (four chapters for Part II). In contrast, the narrative literature section only has one chapter in between.

Part I focuses on insects as individuals and groups in a broad—not strictly scientific—sense, such as the silkworm, cicada, mayfly, click-beetle, firefly, moth, mantis, spider, ant, bee, butterfly, cricket, grasshopper, locust,

5 There are some conferences and workshops devoted to the study of inserts and other invertebrate animals.

6 Wilt L. Idema, *Insects in Chinese Literature: A Study and Anthology*, Cambria Sinophone World Series (Amherst, New York: Cambria, 2019). Interestingly, another important book on animals in Chinese history appeared in 2019, which is *Animals Through Chinese History. Earliest Times to 1911*, edited by Roel Sterckx, Martina Siebert, and Dagmar Schäfer, which I will discuss in the next section.

fly, mosquito, scorpion, louse, flea, and bedbug. Given the myriad species on our planet, these insects only represent a small portion in nature. Yet, in terms of their representations in Chinese literature, and especially in contemporary scholarship, Idema's book already offers a comprehensive account for many insects. This group of insects seems to be closely connected to human life than many other insects, which is why they frequently appeared in premodern Chinese authors' writings. First nine chapters of Part I compiled and translated numerous excerpts from the tradition of *Belles Lettres*, including various genres such as poems (*shi*), rhapsodies (*fu*), lyrics (*ci*), arias (*qu*), and essays (*wen*), chronologically arranged from the earliest sources to the most present. Each chapter has a short and concise introduction explaining literary and cultural backgrounds about the selected literary texts, and its main section is the translations of excerpts. These brief introductions might not be adequate from the perspective of cultural history, but they serve as excellent guides for reading translations. For instance, Chapter 1 deals with the silkworm, perhaps the most important worm for the sericulture in Chinese history. Idema introduced how it was represented in some classical accounts such as the *Book of Poems* and the *Book of Xunzi*, as well as medieval poems, but omitted many studies on cultural history. For instance, Stuart Young noted how medieval Chinese Buddhism treated silkworms. Idema did not rely on Angela Sheng's solid research on the silk as currency on the Silk Road. Instead, he briefly cited the popular book on the Silk Road by Valerie Hansen. Chapter 5 touches on the bee, which should be read together with David Pattinson's chapter on honeybee in *Animals through Chinese History*. While Chapter 6 introduced the locust in Chinese literature, it should remind that many Chinese and Japanese scholars, including myself, also discusses the implications of the "locust disaster" and the locust-eliminating movements in the history of Chinese agriculture. In general, readers might also be interested in reading Ann Heirman's article "How to Deal with Dangerous and Annoying Animals: A Vinaya Perspective" (*Religions* 2019, 10, 113) with Idema's book together.

Nevertheless, I should highlight two strengths in Part I. Idema's mastery of contemporary Chinese scholarship is manifested by his often picking up the most reliable critical editions of these Chinese texts. His erudition acquired from his experience of reading Chinese literature for several decades by annotating numerous historical and literary allusions along with his elegant translations is equally, if not more impressive. Although he did not give sources or citations for most allusions, non-Sinologists readers would be able to rapidly learn about or at least be aware of these stories in the vast literary and cultural traditions of China in more than two millenniums.

In contrast to some excerpts in Part I, such as poems by famous writers that have been known to readers or even already have English translations, Part II includes mostly local popular literary texts that are often known to readers by their collectors, compilers, or performers rather than original authors. Chapters 11–14 are organized with eventful themes, including weddings, funerals, battles and wars, and disputes and court cases. In these events, insects appeared as anthropomorphized figures acting as dramatic roles who could talk and behave. For example, Chapter 11 translated three texts, including *The Precious Scroll on the Marriage of the Mantis*, *The Dragonfly's Abduction of the Bride*, and *The Dung Beetle Abducts His Bride*, with a brief introduction. Since these texts mentioned Bodhisattva and King Yama, they might be connected with the Buddhist thoughts of reincarnations and merit-making, if their Buddhist context could be further explored. Chapter 12 translated two texts focusing on the funerals, such as *The Hundred-Day Insects* and *The War of Insects*. Chapter 13 continues to translate one more Ballad titled *The War of Insects* from Yangzhou and a very lengthy Minnanese ballad from Taiwan titled *The Song of the War of the Fly against the Mosquito*. The last chapter translated three texts focusing on court disputes with extensive conversations, including *The Southern Window Dream*, *The Louse Cries Out His Grievances*, and *The White Louse Voices His Grievances*.

In the conclusion section, Idema compared the insects in Chinese and Western literature. He noted that the silkworm seemed to be absent in premodern Western literature. However, the cicada played an important role in both Chinese and Western literature. Compared to Chinese literature, in classical Western literature, social insects such as bees and ants seem to be the most important insects; and they continued to dominate the Western literary tradition till the modern period. Furthermore, Idema suggested that noxious insects and body parasites seem to be more common in Western literature than in Chinese literature. Yet in Chinese literature, spider appeared as the most erotic insect. It might be interesting if looking into the roles of these different insects in the social life and natural environment in the historical West and China. Another issue might be the difference between insects in North and South China because of different climate and natural environment. After all, the daily experience and ancient knowledge about these insects seem crucial for the literary authors to compose their works.

Overall, this is a fascinating and reader-friendly anthology. Occasionally, there are minor problems, including incorrect translations and untranslated titles. Let me raise several examples from the first and last chapters. In Chapter 1, in the translation of the Origin of the Silkworm by Gan Bao, the opening line in the third paragraph was translated as "The father was exceptionally fond of the animal and provided it amply with food

為畜生有非常之情,故厚加芻養" (p. 29), which seems to be a misunderstanding of the original sentence. It could be translated as "For this animal has extraordinary emotions (toward humans), her father provided it amply with food." In the same text, in the second last line of the last paragraph, the Chinese Character *sang* 喪 was translated as "to bury" (p. 30), which should be "to lose." And in chapter 14, *The Drunken Man's Talk* (*Zuiweng tanlu* 醉翁談錄) could be changed to *The Record of a Drunken Elder's Talk* (p. 225). A folk performing art *Liulin Sanxianshu* 柳林三弦書 remains untranslated (p. 229). It could be translated as *The Three-String Book from the Willow Forest.* They could be easily fixed. Of course, most translations are as much reliable and elegant as they are expected. I believe that readers will enjoy reading this book tremendously.

Animals Studies in Asian History

Animals through Chinese History. Earliest Times to 1911 consists of twelve chapters and offers a comprehensive picture of the knowledge and knowledge practices around animals in Chinese history.[7] As the editors state in the introduction, this volume aims to combine both material culture, especially archaeological findings, and texts to use animals to think about the natural change over the longue durée, as well as to focus on the spiritual and physical roles of animals in society, the state and thought. For me, it is exemplary in taking innovative approaches and offering sophisticated methodologies on many fronts, by exploring and analyzing very rich primary sources, with the exception of visual materials, and it goes beyond conventional agricultural and environmental history. These innovations secure its unique spot in any reading list on Chinese history.

This volume focuses on real animals, or nonhuman animals, and especially on those animals that play the most important roles in human political, economic, and social life, such as pigs, cats, bees, cows, and so forth. Given its emphasis on the knowledge and knowledge practices around real animals in traditional China, it can be regarded as a preliminary volume to supplement Joseph Needham's monumental enterprise *A History of Chinese Science and Civilisation*, in which a volume on animals from the perspective of the history of science and technology has never been published. However, this volume pays less attention to the imagined animals in the spiritual and psychological experience of humans in traditional China.

---

7 Roel Sterckx, Martina Siebert, and Dagmar Schäfer eds., *Animals through Chinese History: Earliest Times to 1911* (Cambridge: Cambridge University Press, 2019).

The twelve chapters in this volume are chronologically arranged, but none is limited to a specific dynasty. Chapter 1, by Adam Schwartz, analyzes the importance of animals to the sacrificial rites for ancestors and natural powers in Shang China and argues that cattle, sheep, and pigs played a complex role in the early development and transmission of Chinese ritual culture, scribal practice, and social interaction by examining material culture, oracle bones, and associated texts from Huayuan zhuang. Chapter 2, by Roel Sterckx, reveals that a sophisticated system was developed for livestock management and professional specialists appeared for preparing animals for ritual purposes in ancient China. These specialists were involved in breeding, selecting, decorating, killing, cutting, drowning, burning, and offering animals following specific procedures. Chapter 3, by Keith Knapp, illustrates that both animals and humans share the same filial piety as the heavenly endowment which blurs the human–animal distinction by examining virtuous animals as moral exemplars in the Confucian stories of filial piety, especially the *Xiaozi zhuan* 孝子傳, from the first to sixth centuries, while connecting animal filiality with knowledge of the medieval natural world. Knapp's other significant contribution is his findings of two differences between animals' and humans' filial piety, including the maternity and reciprocity of animal filiality. I would add that Michael Nylan's recent article "Humans as Animals and Things in Pre-Buddhist China" (*Religions* 2019, 10, 360) perfectly enhances Knapp's discussion.

Chapters 4 and 5 focus on two particular animals: the cat and the bee. Chapter 4, by Timothy H. Barrett and Mark Strange, provides a concise yet thorough history of cats through Chinese history from ancient times to the twentieth century and addresses many crucial issues centered on the practical and symbolic roles and functions of cats in the domestic, religious, cultural, and daily lives of Chinese people. This chapter should be read together with Barrett's two earlier articles on cats in Chinese Buddhism. Chapter 5, by David Pattinson, traces the transformation of Chinese knowledge about honeybees in texts over the centuries and argues that in early texts the taxonomies were vague and the Chinese name for the bee, "feng 蜂," was associated with violence, cruelty, and bad luck. But in the early medieval period, Guo Pu transformed "feng" into a positive, and later in the Song-Yuan periods, "feng" was granted Confucian values of duty, order, and ritual in literary texts.

Chapters 6 and 7 concentrate on texts. Chapter 6, by Francesca Bray, analyzes the presence and absence of farm animals by drawing upon four Chinese agricultural treatises from the sixth to the eighteenth century. The composition of these texts is produced by the entanglement of politics, morality, and elite mores. For this reason, horses and sheep are recorded more prominently than cattle and pigs. Chapter 7, by Martina Siebert, ventures into the relation between the scholar and the animal by focusing on "pulu 譜錄" in the Song Dynasty, a popular scholarly genre in the form of treatise and list dealing with particular animals and material objects. Siebert notes the division between "remote" animals, or the animals constructed from other texts and hearsay, and "close" animals, or those constructed from actual observations.

Chapters 9 and 11 deal with the relations between the state and animals by focusing on the care and management of animals. Chapter 9, by Dagmar Schäfer and Han Yi, explores how the literati thought about systems, function, infrastructure, organization, and processes of farming, managing, and knowing large livestock during the Song Dynasty, and it concludes that central state power shaped the farming of cattle and horses and developed the veterinary sector. Trade, ritual, and moral behaviors played their role in the process of caring for and managing animals. Chapter 11, by Sare Aricanli, discusses the care and management of imperial horses from the linguistic and institutional perspectives by focusing on Mongolian, Manchu, and Han Chinese vocabulary, and it reveals the fluid boundaries between institutions, human and animal medicine, and practices in the center and the steppe in the Qing Dynasty.

Chapter 10, by Vincent Goosaert, explores how late Qing elites wrote about caring for and killing animals in morality books. He contextualizes the early modern views of protecting life in which the boundaries between animals and humans blurred and the protection of life has its social, political, and economic foundations. Chapter 8 by Zheng Xinxian 鄭心閒 on Qianlong's knowledge of animals through classical learning; it suggests that animals played a crucial role in the political life of eighteenth-century China. The last chapter, by Mindi Schneider, examines the economic importance of the pork industry by looking into its global–local and past–present dynamics in modern and contemporary Chinese cultural and political contexts.

Overall, this volume is a very welcome addition to the current scholarship on the animals or human–animal relations in Chinese history, as well as to animal studies in general.

***

Many books have been written about the giant panda, undoubtedly one of the most popular—and often considered one of the cutest—animals on Earth. E. Elena Songster's new book *Panda Nation: The Construction and Conservation of China's Modern Icon* presents the first comprehensive ethnographic account of how the giant panda became a national icon in the context of the rise of the People's Republic of China (PRC) and how this iconic image contributed to the evolution of the identity of the PRC in the international community.[8] In pointing to the book's comprehensiveness, I mean that it encompasses the exercise of state power in the PRC, various historical actors, the giant panda itself, panda reserves, the other animals inhabiting those reserves, the wider environment, and the role of other international players. This combination secures for this book a significant place in contemporary scholarship. It is also noteworthy that the author keeps the book to a moderate length while still addressing political power, human–animal relations, and environmental and ecological concerns.

This book is very impressive in successfully weaving a broad and diverse range of sources into a grand narrative about the giant panda as a precious and rare species, a scientific object, and a political symbol in the history of the PRC. As the author notes, she has drawn on "Chinese- and English-language sources from local, provincial, and national archives, newspapers, scientific articles, governmental policies, surveys, books, and laws, as well as insights from experts, officials, and zookeepers." She also "analyzes artwork, advertisements, maps, photographs, and statistical data, as well as first-hand observations of the pandas and the habitat itself." (p. 3) Research must have required a painstaking search for sources across genres. Taken together, they illuminate the exercise of PRC state power at the local and central levels.

Aside from a brief introduction outlining the structure of the book and the major research questions and a conclusion reaffirming some of the main arguments, the eight central chapters of the book are arranged in roughly chronological order, with each chapter focusing on one or two particular issues. Chapter 1 offers a brief history of the scientific discoveries and taxonomic debates relating to the giant panda and its fossils beginning in the year 1869, when Fr. David, a French-Catholic missionary, introduced it to the West. From the perspective of the history of science, this chapter argues that the scientific study of the giant panda, and its exclusive existence in China, motivated the PRC to present this animal as a national symbol. Chapter 2 does not directly focus on the giant panda, but rather offers historical, political, and economic background for the creation of nature reserves for protecting precious and rare animals in the early years of the PRC. It examines the process of establishing Mt. Dinghu the PRC's first nature reserve in 1956 under the influence of Soviet experts. The author suggests

8 E. Elena Songster, *Panda Nation: The Construction and Conservation of China's Modern Icon* (New York: Oxford University Press, 2018).

that interest in natural preservation was initially limited to the scientific community, but that the central government quickly became involved as part of the socialist state-building project of the 1950s and 1960s. This chapter also investigates the implementation of the government policy of "protecting, rearing, hunting" during the Great Leap Forward and the subsequent strategy of economic recovery.

Chapter 3 returns to the giant panda and discusses how "the creation of the Wanglang Panda Protection Reserve in 1965 demonstrated how a concept such as nature-protection policy could originate in the nation's top legislative body and be transformed into physical reality in the remote mountain ranges of southwestern China." (p. 69) Analyzing data from archives, interviews, and fieldwork, this chapter argues that the creation of the giant panda protection reserve in Wanglang incorporated provincial, local, and individual interests. This research illustrates the process by which the central government, local ethnic minority hunters from the Baima community, and individuals such as Zhong Zhaomin worked together to make the reserve a reality. Chapter 4 traces how the giant panda became a national treasure during the Cultural Revolution. Interestingly, while the Cultural Revolution is often thought of as a catastrophe, by analyzing scientific research on the giant panda and the cultural—especially artistic—visions of the panda, the author suggests that, "the Cultural Revolution atmosphere launched the giant panda into the role of national icon" (p. 73).

Chapter 5 discusses the famous practice of "panda diplomacy," but draws particular attention to the profound effects of China's domestic conservation policy on the wild population of the giant panda, which makes it different from many other publications on the subject. Chapter 6 turns to two case studies on the panda rescue movements in 1976 and 1983, and provides a detailed account of how central, provincial, and local government agencies and reserves responded to growing crises of death and disease among the panda population, and eventually how the response to the crisis was transformed into a domestic and international movement to save the giant panda as a Chinese national treasure. The author particularly notes how these movements pushed a number of talented young people to pursue careers in conservation biology and to become specialists focusing on the giant panda (p. 126).

Chapter 7 moves to the reform and opening-up period. First, the author discusses how the Wanglang Reserve for protecting the giant panda became a victim of the flourishing local ethnic ecotourism promoted by the Baima Tibetan community. Next, she gives an account of how the managers of the reserve responded to this problem by transforming it into a cooperative scientific research base open to both domestic and international zoologists, biologists, and ecologists, as well as conservation experts. The final substantive chapter analyzes how the PRC government shifted from a policy of loaning giant pandas out for short-term celebrations such as the Olympic Games to a policy of arranging long-term loans in order to develop scientific research collaboration. It also examines the political debate over the practice of offering pairs of pandas as gifts across the strait to Taiwan.

This book is strong in its sound analysis, fruitful discussion, and convincing arguments. However, it could be improved in several ways. One puzzle for the reader is that the book seems to put too much weight on the uniqueness of the PRC's state building enterprises, such that it risks ignoring some political and cultural continuities from the Late Imperial and Republican Eras. Another puzzle is the complete neglect of established Japanese scholarship, especially Ienaga Masaki's work on panda diplomacy.[9] These omissions weaken some of the claims made in *Panda Nation*.

For instance, in the introduction, the author notes that "The transformation of the giant panda from its unsung and hidden existence in the remote mountains of western China to a position of national and international recognition and influence is in many ways a byproduct of a larger story: the rise of the People's Republic of China." (p. 1) This is hardly the case; before the founding of the PRC, the giant panda had already won the heart of the British in the 1940s due to Jiang Yi's 蔣彝 writings and paintings, as Ienaga notes in detail in his *Panda Diplomacy* (pp. 41–47). Additionally, *Panda Nation* pays little attention to panda diplomacy during the 1940s, when the government of the Republic of China incorporated the giant panda into its grand propaganda project against Japanese invasion, using it to portray China as a peaceful and warm nation deserving of sympathy and support from other nations. In its claim that the panda became an expression of nationalism, a tool for diplomacy, and a means for international cooperation and scientific exchange (p. 5), *Panda Nation* also underestimates the panda's place as the symbol of Chinese culture. This argument omits the role of Jiang Yi's work in England, which painted the giant panda as a symbol of traditional Chinese culture.

Because *Panda Nation* focuses on the PRC, its chapter on panda diplomacy is particularly weak, in that it does not offer sufficient discussion of the international relations between the PRC, Soviet Russia, Britain, and

---

9 Ienaga Masaki 家永真幸, *Panda gaikō* パンダ外交 (Panda Diplomacy), Tōkyō: Mediafakutorī, 2011; and chapter 6 of his new book *Kokuhō no seijishi: "Chūgoku" no Kokyū to panda* 国宝の政治史: "中国"の故宮とパンダ (The Political History of Natural Treasures: China's Palace Museum and Panda) (Tōkyō: Tōkyō Daigaku Shuppankai, 2017), 211–49.

Japan in the second half of the twentieth century. Ienaga's book offers a much better account, including interesting details of how Britain obtained a giant panda from a private business operator in Australia due to the PRC's ban on exporting pandas to enemy nations, and how Britain and Soviet Russia collaborated to mate their pandas in the 1960s (albeit failing to produce a baby), even though they belonged to opposing political and military camps. *Panda Nation* also fails to examine how panda policy related to the competition for legitimacy between the PRC and ROC. In the 1960s, China developed its nature protection policy in response to the challenge of the ROC government in Taiwan.

Occasionally, *Panda Nation* cites unreliable secondary sources. For example, citing Shuping Yao's article on the history of the Chinese Academy of Sciences 中國科學院 (CAS), the book states that, "After the founding of the PRC, the new government saw science as a means of both liberating and strengthening China as a nation. From the first month of its rule, it actively organized such scientific institutions as the Chinese Academy of Sciences, which governed six research institutions plus the Institute of History and languages (p. 28)." This is not the case. In 1950, when the PRC founded the CAS, only the Institute of Archaeology 考古所 and the Third Institute of History 第三歷史研究所 were established. The Third Institute was developed out of the Teaching Unit of the North China People's Revolutionary University 華北人民革命大學 under the leadership of Marxist historian Fan Wenlan 范文瀾 and focused on modern Chinese history. The planned First Institute 第一歷史研究所 (focusing on ancient history) and Second Institute 第二歷史研究所 (focusing on medieval history) were founded later, in 1954.

Although there are some minor weaknesses, *Panda Nation* makes a crucial contribution to our understanding of science and the PRC nature conservation enterprise. Some of the book's ideas are inspiring and well worth pursuing further. While discussing the formation process of the nature conservation policy in the early years of the PRC, the author notes that PRC articulations of state-sponsored nature conservation were very similar to those in Progressive Era America, concluding that "Communist conservation and capitalist conservation shared the same basic rationale for protecting natural resources and employing scientific methods to do so. These parallels become more understandable in light of the fact that Zhu Kezhen 竺可楨 and other contemporary leading scientists in China were educated in the United States in the early twentieth century." (pp. 39–40) In fact, many scientists and engineers including Zhu Kezhen were educated at Tsinghua and then moved to study in the United States in the 1910s and 1920s, leading many of them to accept Progressive Era ideas about social engineering.

Another interesting point made in *Panda Nation* is that during the Cultural Revolution conservation biology benefited from the giant panda's image as a national treasure. This can be understood in a broader context, noted by many scholars, in which the nationalist spirit of the 1960s also motivated many other scientific achievements in fields as diverse as rocket science, nuclear technology, biology, mathematics, and medicine (including the synthesis of crystalline bovine insulin in 1965 and the discovery of artemisinin, which won a Nobel Prize for Tu Youyou 屠呦呦 decades later).

Despite some minor shortcomings, this book is very welcome as a work of scholarship contributing to the study of the PRC. Its accessible style makes it an excellent choice for undergraduate courses on the history of the PRC, science and society, environmental history, and the history of animals.

Animal/Plant Studies and Asian Medical History

Based on his revised dissertation, Liu Yan's new book *Healing with Poisons: Potent Medicines in Medieval China* is a welcoming addition to the English-language scholarship on the history of medicines in China, focusing on the medieval transformation of poisons as medicines.[10] In the past two decades, the history of pharmacology, pharmacy, and medicines in China has experienced a booming development across the globe. Many books focus on early modern, modern, and contemporary periods. For example, just in the past couple of years, we have seen the publications of *Know Your Remedies: Pharmacy and Culture in Early Modern China* (Princeton, 2020) and *Mao's Bestiary: Medicinal Animals and Modern China* (Duke, 2021). However, many books on premodern periods have been published in Chinese, Japanese, and French, as Liu Yan also noted in the introduction of his new book. On the one hand, these books resulted from the flourishing of cultural history that focused on the body, health, medicine, and life. On the other hand, there was also inspiration from newly available materials, such as manuscripts found in Dunhuang and other sites in Central Asia and entombed stone inscriptions. Indeed, Chinese and Japanese scholars have continued the tradition of studying materia medica (bencao 本草) to compile, edit, and study these manuscripts. In recent years, some East Asian scholars also attempted to incorporate new concepts to interpret these new materials in light of the history of medicine and material culture. One of the strengths of Liu's book is to digest numerous secondary sources in East Asian languages.

10 Yan Liu, *Healing with Poisons: Potent Medicines in Medieval China* (Seattle: University of Washington Press, 2021).

Besides incorporating secondary sources, *Healing with Poisons* focused on two genres of texts as primary sources, meteria medica (*bencao*), and formula books (*fangshu* 方術). I would further point out that from the perspective of material culture since there were three major material sources for Chinese medicines in medieval China: animals, plants, and minerals, many texts on the roles of plants, animals, and minerals in Chinese medical history might not be categorized into the two genres noted in *Healing with Poisons*, so their values for this theme might have been underestimated. For example, the text on zoomancy or animal divination collected in *Treatise on the Auspicious Signs of Heaven and Earth* (Tiandi ruixiang zhi 天地瑞祥志) often mentioned the animals serving as medicine for healing illness.[11]

Healing with Poisons has three parts and seven chapters. The first two parts trace the origins and evolving transformation of the "*du* 毒" as "potent" or "potency" from poison to medicine in the textual sources from the Han to the Tang Dynasties. And the second part focuses on the relationship of state power to pharmaceutical knowledge and its medical and political implications. The third part turned to the role of the poison in life enhancement. The conclusion section summarizes the discussions and makes several arguments. First, Liu argues that the traditional Chinese pharmacy encompassed various substances possessing great transformative capacity. Second, he argues that the materiality of Chinese drugs was fluid, and the medical knowledge was transformative across political and social spaces. Third, he also argues that the medicines could alter the body for both positive and negative functions. Fourth, Liu suggests that the center of medical activities was shifted to the northwestern region in the seventh century due to the establishment of state power. The state and the elite class played active roles in producing, regulating, and institutionalizing medical knowledge.

In particular, Part I has two chapters. Chapter 1 traces the changing meanings of "*du*" in early Chinese sources, from the earliest materia medica, *The Divine Farmer's Classics*, to Tao Hongjin's 陶弘景 *Collected Annotations.* He argued that the "*du*" was more sophisticated than simply the negative connotation of "poison" in early Chinese texts, which denoted the strong sense of potency for a medicine's therapeutic power, and the role of "*du*" in healing was established in the sixth century. He also suggested that political visions and cosmological views contributed to the blurry boundary between poison and medicine due to the Yin-Yang dynamism. Chapter 2 analyzes the technical issues involving the "*du*" as medicine, such as dosage control, drug combination, processing aconite, and the drug trade, and concludes with the fluid materiality of Chinese medicines.

11 For more details about this book, see my discussion in Chapter 4 of this book.

Part II consists of three chapters. Chapter 3 examines the medieval concepts of demonic infestation (guizhu 鬼注), poison (*gu* 蠱), and witchcraft (*gu*) in the context of the Sui and Tang political history. It traces the idea that in the Han dynasty, the idea of disquieted demonic spirits as the sources of hauntings, and that "zhu 注" (pouring) referred to the transmission of illness through the dead body as the source of contagious pollution, which led to the epidemics. Then early medieval Daoist tradition developed many techniques for harmonizing these restless spirits. However, Sun Simiao 孫思邈 developed formulas for using potent drugs to drive the sinister demons out of the infested bodies, which became a method of "fighting poison with poison." Then this chapter continues to discuss the *gu* vermin and its implication of the seductive power of women. Using the cat demon in the political history of the Tang dynasty as a case study, Liu illustrates that in the Tang dynasty, the state power aimed to expel the *gu* practitioners to stabilize the social order, which replaced the old discourse developed in the Han dynasty when medical writing focused on the resonance between the individual, the state, and the cosmos.

Chapter 4 offers a concise institutional history of medicines and medical texts, particularly *Newly Revised Materia Medica* 新修本草 in the Tang dynasty, and analyzes the local medical knowledge using the manuscript sources from Dunhuang. However, the section exploring the manuscript from Dunhuang seems less convincing and insufficient. Numerous primary sources have been found, and there are many secondary sources on this topic in Chinese, Japanese, and Western languages. In terms of the institutional history, I find that there might be a parallel development of medical history and the history of divination in ancient and medieval China since there was no clear boundary between physicians and diviners in ancient China. Many aristocratic families possessed their divination knowledge and techniques since the Han dynasty. Still, in the medieval periods, the state also attempted to play more active roles in regulating the divination knowledge and techniques by institutionalizing them.[12]

Chapter 5 is a case study on Sun Simiao and his Essential Formulas to show the importance of the formula books in the study of medieval medicines and medical practice. It briefly mentioned Sun's connection with Buddhism. As I noted earlier, there were more genres of medical texts in medieval China beyond that of materia medica and formula books. My earlier study indicated that Daoxuan 道宣, a Vinaya master in the Tang dynasty, actually listed three crucial categories of materials that could be offered to the monastic

12 See Chapter 4.

community: healing devices, medical texts, and medicines. In the category of medical texts, he listed many genres of texts such as materia medica (*bencao*), various formula books (*zhufang* 諸方), classics of the pulse (*mojing* 脈經), and the methods of flowing and blocking from the Bright Hall (mingtang liuzhu 明堂流注), etc. Of course, Chapter 5 focused on medicines, so it naturally omitted other traditional Chinese medical technology such as acupuncture. However, combining this chapter with Chapter 4 on the institutional history of medicines and medical texts, we see a crucial discussion how various medical texts were categorized by both the state power and the religious (both Buddhist and Daoist) monastic community. For instance, the "treatise of classics and texts" (Jingji zhi 經籍誌) of the *History of the Sui Dynasty* offered an example of the state institutionalization of texts since the *History of the Sui Dynasty* was an official history sponsored by the court and overseen by the prime minister.[13]

The last part has two chapters dealing with enhancing life by taking drugs. Chapter 6 focuses on mineral ingredients such as the five-stone powder and its function for revitalizing the body and illuminating the mind. Chapter 7 deals with taking elixir and its position in medieval Chinese alchemy.

Overall, *Healing with Poisons* is an exciting and concise book that shows a sophisticated picture of Chinese drugs in the medieval period. Besides the author's promise of further exploring the medical writings of Tang scholars, I also wonder about the contributions of non-Han cultures surrounding China in the medieval period. Numerous Indo-Iranian, Tibetan, and Turkic-Uyghur physicians, diviners, and political and religious agents deal with drugs and potency on the Silk Roads and they greatly changed the landscape of medical knowledge and techniques in China. Is it possible to ask if there was a global network of drugs in the medieval periods? These questions delve beyond the scope of this book. Still, I hope the readers keep in mind that medieval China was very much open to the influence of various cultures beyond Dunhuang and was impacted by neighboring cultures in East Asia.

***

Liz P. Y. Chee's new book *Mao's Bestiary: Medicinal Animals and Modern China* is an exciting addition to the current scholarship on the history of the PRC.[14]

---

13 Huaiyu Chen, "A Preliminary Study on a Case of the Exchange of Medical Knowledge between Daoxuan and Sun Simiao (Daoxuan yu Sun Simiao yixue jiaoliu zhi yizheng lice 道宣與孫思邈醫學交流之一證蠡測)," in *Dunhuang Tulufan yanjiu* 敦煌吐魯番研究 9 (2006): 403–8.

14 Liz P. Y. Chee, *Mao's Bestiary: Medicinal Animals and Modern China* (Durham, NC: Duke University Press, 2021).

This book focuses on how various social agents, including the state, the industry, and the researchers, handled animals as materia medica from the 1950s to the 1980s. This important study offers an excellent interdisciplinary study on the impacts of political rhetoric and state policy on the development of Chinese animal-based medicine, which is manifested by Chee's insightful analysis of numerous Maoist slogans. From a broad historical context, Chee reveals multilayer tensions between Chinese patriotism and colonialism, the state policy and local practice, the central planning economy and the market economy, and the economic interest and animal bioethics, based on the crucial information she painstakingly mined from various medical journals and magazines, especially those preserved at Guangzhou University of Chinese Medicine. In examining the intermingled relations among politics, economy, and science, Chee documents the evolving development of animal-based pharmaceuticals in contemporary China. This new book echoes the growing scholarly interests in animal studies and medical history, as well as the history of the PRC.

Besides the "introduction" and "conclusion" sections, this book's five bulk chapters generally follow chronological order to address some critical issues in five historical periods. The "Introduction" section introduces research methods and sources and clarifies some critical terms and phrases in a contemporary Chinese context. Chapter 1 focuses on the period from 1950 to 1957 and examines how the PRC transformed Chinese pharmaceutical production by abandoning Chinese medicine and retaining Chinese drugs and merging the drug-making sector into a state-controlled central planning system as part of the state-building project. This chapter argues that, though the separating medicine from Chinese drugs can be traced back to the Republican era, the PRC consolidated traditional medicine making business into the state-controlled factory system in the name of "scientizing Chinese medicine" through introducing science-based innovation for producing these drugs and training a new generation of pharmacists and technicians. However, in this era, animal-based medicine was not the focus.

Chapter 2 analyzes the influence of the Soviet Union on the animal-based drug-making business in China. It suggests that China accepted the Soviet "tissue therapy" for a couple of reasons: technically, both China and Russia shared the practical interest in indigenous plants and animals as raw medicinal materials for their shared borders and environment, such as ginseng and deer antler; economically, both China and Russia faced the sanctions of Western medicine from America and Western Europe, so they turned to identify available plant and animal resources for self-sufficiency in drug making since China and Russia have large rural populations who rely on local medical resources.

Chapter 3 examines the rise of medicinal animal framing during the Great Leap Forward era. It argues that during the Great Leap Forward era, China was ambitious to become self-sufficient in drug making and export drugs to the world for surpassing Britain. The Chinese government launched the movement to discover new animals as raw medicinal materials and innovate to make drugs from these animals. So the state expanded the list of animals as medicinal materials, including insects, small reptiles, and seahorses, which led to the large-scale movement of faunal medicalization. With the strong demand from the exporting market, China expanded its musk deer farming business and developed a farm-based fish oil industry.

Chapter 4 explores the issue of folk remedies during the Cultural Revolution, with a detailed case study of chicken blood therapy. It demonstrates an exciting case that Chinese doctors and medical professionals innovatively "scientized the chick blook therapy" by modern biomedical technology such as lab testing and injection method for legitimating and promoting this new therapy among the mass, which particularly attracted the population in the rural area.

Chapter 5 uses the development of bear farming as a case study to examine how the reform policy in Deng Xiaoping's era marketized the animal-based drug industry. Animal farming not only served the pharmaceutical industry but also benefited the health supplements making business. The marketing economy transformed these animals into agricultural resources. The bear farming business brings together many interest groups, including bear bile producers, dealers, scientists, technicians, and farmers, in the global context in which even Japanese scientific experience on bear bile study was introduced.

In "conclusion" section, the author argues that the animal-based pharmaceutical industry sustained and continued today for two reasons: folk remedies for discovering the medicinal use of animal parts and tissues as supplements to modern biomedicine, the economic profits of exporting animal-based drugs for the state. However, animal farming in China has drawn concerns and criticism from the activists of bioethics and animal welfares and scientists since the 1980s.

Chee's new book is well structured and clearly written, which brings up a very joyful reading. The lucid style will be welcomed by readers from various academic backgrounds. The sources in this book could be improved a little bit. In particular, Chapter 2 mostly used sources from the China-published journal articles and news reports and English-language secondary scholarship on Soviet medicine, but no secondary sources in Russian. Overall, this book only cited several news reports in Japanese but ignored many secondary resources in Japanese scholarship. Many Japanese-language journals, such as *Kampo Medicine* and *Farumashia*, published numerous articles on traditional Chinese herb and animal-based medicines. Nevertheless,

I would recommend this exciting book to anyone who is interested in animal studies, Chinese history, and the history of medicine.

Concluding Remarks

While looking forward to the prospective scholarship in the future, I would propose 3-T approaches, including trans-species, transnational, and translingual studies. In other words, future scholarship should cross the boundaries of species, nations, and languages in terms of themes, regions, and sources.

First, while dealing with animals, there are always blurry boundaries between animals and humans. Chapter 6 of the current book offers a concrete case study on the were-tiger in Asian culture, which illustrates the trans-species relations. Local literature and folklore created images of the were-tigers in East and Southeast Asian regions. They were similar to the werewolves in Europe and elsewhere, transforming their body forms but sharing common emotions and cognitions with humans. The situation was even more sophisticated in Chapter 5 on the twelve zodiac animals since the zodiac animals could be portrayed as both Bodhisattvas and demons, depending on the cultural contexts.

Furthermore, another significant issue from the perspective of trans-species studies is the zoonoses and medicine. In the human history, some trans-continental plagues such as the black death swiftly spread and devastated human society. Some small communities were even entirely wiped out by the plagues. Contemporary scholarship has developed a new concept of "One Health, One Medicine, and One History" for bringing an integrated approach to diseases and plagues across the boundaries between humans and animals.[15] The current COVID-19 pandemic again reminds us of the significance of understanding the zoonotic issue in a more comprehensive

15 Anne Hardy, "Animals, Disease and Man: Making Connections," *Perspectives in Biology and Medicine* 46 (2003): 200–15; Robert G. W. Kirk and Michael Worboys, "Medicine and Species: One Medicine, One History?" in *The Oxford Handbook of the History of Medicine*, ed. Mark Jackson (Oxford: Oxford University Press, 2011), 561–77; Jakob Zinsstag, Esther Schelling, David Waltner-Toews, Maxine Whittaker, and Marcel Tanner eds., *One Health: The Theory and Practice of Integrated Health Approaches* (Wallingford: CAB International, 2015); Abigail Woods, Michael Bresalier, Angela Cassidy, and Rachel Mason Dentinger eds., *Animals and the Shaping of Modern Medicine: One Health and Its Histories* (London: Palgrave Macmillan, 2018); Frank M. Snowden, *Epidemics and Society: From the Black Death to the Present* (New Haven: Yale University Press, 2019); Christos Lynteris ed., *Framing Animals as Epidemic Villains: Histories of Non-Human Disease Vectors* (Cham: Palgrave Macmillan, 2019).

cultural, environmental, and social context. The increasing activities of humans toward natural wildness and the introduction of wild animals into human economic life paved the way for transmitting diseases from wild animals to humans. The climate change might lead the melt of glacier and release the frozen virus.[16]

Second, except for Ohnuma's book, most books covered in the above discussion focused on animals in China. However, it is worth noting that the study of animals in China is always a multidimensional field. Many of these studies display a strong transnational and cross-cultural orientation. Ohnuma's book reminds us of many Jātaka stories that also appeared in the Chinese translations and might impact Chinese literature. Some themes she dealt with in her book *Unfortunate Destiny* continued and transformed Chinese literature. Chapter 5, "Changing Images of Buddhist Zodiac Animals in Medieval Chinese Buddhist Literature," in my current book demonstrates that medieval Chinese Buddhist commentators would modify the narratives and interpretations while dealing with the themes in early Chinese translations of Buddhist texts to meet the needs of the Chinese readership and audience. So they introduced the indigenous Yin-Yang and Five-phase theories to reorder the twelve zodiac animals and eventually expand the list to accommodate more animals. Some texts appearing in Idema's book also indicate the Buddhist background from the early period. For example, Idema mentioned in the Dunhuang manuscripts of *Rhapsody on the Swallow* (Yanzi fu) that the court case was brought to King Yama. The latter ruled the underground world,[17] which was, of course, from the Buddhist tradition.

Chapter 6 of this current book also shows that some stories in medieval Chinese Buddhist literature also appeared in Korea and were rewritten by Korean authors and editors. In the first two chapters on plants, I briefly mentioned several exotic plants from other parts of Asia that were introduced to China along the Silk Road. A similar case could be applied to animals. Various ethnic groups from Central and South Asia and Southeast Asia introduced numerous animals as tributes and companions to East Asia. Many of these exotic animals played critical roles in shaping the history of Asia, which deserves to be scrutinized. Animal diplomacy can grow up to become an independent field, given numerous cases among various empires and regimes in Asian history. The panda diplomacy in contemporary history was even more fascinating. A global history of panda diplomacy from

16 Zhong, Z. P., Tian, F., Roux, S. et al., "Glacier Ice Archives Nearly 15,000-Year-Old Microbes and Phages," *Microbiome* 9, no. 160 (2021). https://doi.org/10.1186/s40168-021-01106-w.

17 Idema, *Insects in Chinse Literature*, 227–28.

the 1940s to today is still missing in the current historiography. Therefore, the transnational and global perspective would considerably offer us a broad picture of animals across the globe.[18]

This trans-Asian and global perspective is also closely connected with the third approach: the translingual one. It has two dimensions, including bringing up various voices in multilingual sources while studying animals and tracing the spread and transmissions of animals and their implications across the boundaries of ethnic groups who spoke different languages. While reading Ohnuma's study on animals in Indian Buddhism that focused on indigenous sources in Sanskrit, Pali, and Prakrit languages, I was motivated to think about some paralleled Buddhist sources that survived in early Chinese translations since some ideas on animals in these paralleled sources might be reintroduced to the Chinese readers to accommodate the local knowledge in medieval China and particular intellectual and spiritual needs.

Furthermore, explorers, archaeologists, and collectors have discovered and collected many manuscripts and inscriptions from archaeological sites and ancient ruins along the Silk Road. Many of these manuscripts and inscriptions from the long history of Asia were written in many different languages, such as Khotanese, Middle Persian, Mongolian, Sogdian, Tangut, Tibetan, Tocharian, and Turkic-Uyghur. Animal studies based on these manuscripts and inscriptions in ancient Asian languages remain underdeveloped. Further exploration will help us gain a more comprehensive understanding of animal culture in Asia and its role in shaping Asian history and beyond.

18 Annika A. Culver, *Japan's Empire of Birds: Aristocrats, Anglo-Americans, and Transwar Ornithology* (London: Bloomsbury, 2022). It examines the ornithological practices of Japanese explorer-scientists in the Japanese empire and how the ornithological knowledge circulated between Japanese and British/American spaces in the transwar period from the 1920s to the 1950s.

BIBLIOGRAPHY

Primary Sources

Abbreviations

T = *Taishō shinshū Daizōkyō* 大正新脩大藏經. Edited by Takakusu Junjirō 高楠順次郎 and Watanabe Kaigyoku 渡邊海旭. Tokyo: Taishō Issaikyō Kankōkai, 1924–1929.

X = *Manji shinsan zoku zōkyō* 卍新纂續藏經. Tokyo: Kokusho Kankokai, 1905–1912.

Beimeng suoyan 北夢瑣言. By Sun Guangxian 孫光憲. Shanghai: Shanghai guji chubanshe, 1981.

Biya 埤雅. By Lu Dian 陸佃. Annotated by Wang Minhong 王敏紅. Hangzhou: Zhejiang daxue chubanshe, 2008.

Chen shu 陳書. By Yao Silian et al. Beijing: Zhonghua shuju, 1972.

Chunqiu Zuozhuan zhu 春秋左傳注. By Zuo Qiuming 左丘明. Annotated by Yang Bojun 楊伯峻. Beijing: Zhonghua shuju, 1981.

Da baoji jing 大寶積經 (Skt. *Mahā ratnakūṭa sūtra*). Translated by Bodhiruchi 菩提流志. T. 11, no. 310.

Da fangdeng daji jing 大方等大集經 (Skt. *Mahāvaipulya mahāsamghaṭā sūtra, MMS*). Translated by Dharmakṣema 曇無讖. T. 13, no. 397.

Da fangguangfo huayanjing shu 大方廣佛華嚴經疏. By Chengguan 澄觀. T. 35, no. 1735.

Da fangguang sanjie jing 大方廣三戒經. Translated by Dharmakṣema 曇無讖. T. 11, no. 311.

Đại Việt sử ký toàn thư 大越史記全書. Punctuated and Annotated by Sun Xiao 孫曉. Chongqing: Xinan shifan daxue chubanshe, 2016.

Daixuan shiji 待軒詩記. By Zhang Cizhong 張次仲. Wenyuange Sikuquanshu 文淵閣四庫全書版, vol. 82. Shanghai: Shanghai guji chubanshe, 2003.

Danqian yulu 丹鉛餘錄 (Records of an Exiled Man). By Yang Shen 楊慎. Shanghai: Shanghai guji chubanshe, 1992.

Erya yi 爾雅翼. By Luo Yuan 羅願. Sikuquanshu cunmu congshu 四庫全書存目叢書. Shandong: Qilu shushe, 1995.

Erya Zhushu 爾雅注疏. Annotated under the supervision of Li Xueqin 李學勤. Beijing: Beijing daxue chubanshe, 1999.

Fanwang jing 梵網經 (Skt. *Brahmajāla Sūtra*). Translated by Kumārajīva 鳩摩羅什. T. 24, no. 484.

Fayuan zhulin 法苑珠林. By Daoshi 道世. T. 53, no. 2122.

Fazang suijin lu 法藏碎金錄. By Chao Jiong 晁迥. Dazang jing bubian 大藏經補編, 27, no. 153.

Foshuo chaoriming sanmei jing 佛說超日明三昧經. Translated by Nie Chengyuan 聶承遠. T. 15, no. 638.

Foshuo Da kongque zhou wangjing 佛說大孔雀呪王經. Translated by Yijing 義淨. T. 19, no. 985.
Fozu lidai tongzai 佛祖歷代通載. By Nianchang 念常. T. 49, no. 2036.
Genben shuoyiqieyou bu pinaiye 根本說一切有部毘奈耶 (Skt. *Mūlasarvāstivāda Vinayavibhaṅga*). Translated by Yijing 義淨. T. 23, no. 1442.
Guanding jing 灌頂經. Translated by Srimitra 帛尸梨蜜多羅. T. 21, no. 1331.
Guangyi ji 廣異記. By Dai Fu 戴孚. Beijing: Zhonghua shuju, 1992.
Guben Shanhaijing tushuo 古本山海經圖說. Edited by Ma Changyi 馬昌儀. Ji'nan: Shandong huabao chubanshe, 2001.
Guoqing bailu 國清百錄. By Guanding 灌頂. T. no. 1934, vol. 46.
Guoyu 國語. Attributed to Zuo Qiuming 左丘明, annotated by Wu Guoyi 鄔國義 et al. Shanghai: Shanghai guji chubanshe, 1994.
Hanshu 漢書. Compiled by Ban Gu 班固 (32–92) et al. Beijing: Zhonghua shuju, 1990.
Hou Hanshu 後漢書. Compiled by Fan Ye 范曄 et al. Beijing: Zhonghua shuju, 1999.
Huashu 化書. By Tan Qiao 譚峭. Beijing: Zhonghua shuju, 2009.
Jin shu 晉書. By Fang Xuanling 房玄齡 et al. Beijing: Zhonghua shuju, 1974.
Jin zhongxing shu 晉中興書. By He Fasheng 何法盛. Huang Shi 黃奭 ed. Yangzhou: Jiangsu guangling guji keyinshe, 1984.
Jingci yaoyu 淨慈要語. By Yuanxian 元賢. X. 61, no. 1166.
Jushi zhuan 居士傳. By Peng Jiqing 彭際清. X. 88, no. 1646.
Leishuo jiaozhu 類說校注. Compiled by Zeng Zao 曾慥. Fuzhou: Fujian Renmin chubanshe, 1996.
Liangchu qingzhong yi 量處輕重儀. By Daoxuan 道宣. T. no. 1895, vol. 45.
Lieyizhuan deng wuzhong 列異傳等五種. By Ren Fang 任昉 et al. Beijing: Wenhua yishu chubanshe, 1988.
Lĩnh Nam chích quái deng shiliao sanzhong 嶺南摭怪等史料三種. Edited by Dai Kelai 戴可來, Yang Baoyun 楊保筠, and Wu Shiying 武世營. Zhengzhou: Zhongzhou guji chubanshe, 1991.
Longxing biannian tonglun 隆興編年通論. By Zuxiu 祖琇. X. 75, no. 1512.
Lüshi Chunqiu jiaoshi 呂氏春秋校釋. By Lü Buwei 呂不韋. Annotated by Chen Qiyou 陳奇猷. Shanghai: Xuelin chubanshe, 1990.
Lüxiang gantong zhuan 律相感通傳. By Daoxuan. T. 45, no. 1898.
Maoshi mingwu tushuo 毛詩名物圖說. By Xu Ding 徐鼎. Compiled by Wang Chenglüe 王承略. Beijing: Tsinghua daxue chubanshe, 2006.
Maoshi Zhengyi 毛詩正義. By Kong Yinda 孔穎達. Compiled by Ma Xinmin 馬辛民 and Li Xueqin 李學勤. Beijing: Beijing daxue chubanshe, 2000.
Maoshi Zhengyi 毛詩正義. By Zheng, Xuan 鄭玄. Beijing: Beijing daxue chubanshe, 1999.
Miaofa lianhua Jing maming pusa pin, no. 30 妙法蓮華經馬明菩薩品第三十. Manuscript from Dunhuang. T. 85, no. 2899.
Mingshi 明史. Compiled by Zhang Tingyu 張廷玉 et al. Beijing: Zhonghua shuju, 1974.
Mohe sengzhi lü 摩訶僧祇律 (Skt. *Mahāsāṃghika-Vinaya*). Translated by Buddhabhadra 佛陀跋陀羅 and Faxian 法顯. T. 22, no. 1425.
Mohe zhiguan 摩訶止觀. By Zhiyi 智顗. T. 46, no. 1911.
Qiaozi wuxing zhi 譙子五行志. By Pu Yangxia 濮陽夏. In *Xijian Tangdai tianwen shiliao sanzhong* 稀見唐代天文史料三種. Edited by Gao Keli 高柯立. Beijing: Guojia tushuguan chubanshe, 2011, 585–751.
Quan Tang shi 全唐詩. Compiled by Peng Dingqiu 彭定求 et al. Beijing: Zhonghua shuju, 1960.
Shangshu zhonghou 尚書中候. Annotated by Zheng Xuan 鄭玄. Yangzhou: Guangling shushe, 2008.

Shanhai jing 山海經. Edited by Yuan Ke 袁珂. Chengdu: Bashu shushe, 1992.
She dasheng lun ben 攝大乘論本. Translated by Xuanzang 玄奘. T. 31, no. 1594.
Shengchao poxie ji 聖朝破邪集. By Xia Guiqi 夏瑰琦. Dazangjing bubian 大藏經補編, 28, no. 155.
Shetou jian taizi ershibaxiu jing 舍頭諫太子二十八宿經 (Skt. *Śārdūlakarnāvadāna*). Translated by Dharmarakṣa 竺法護. T. no. 1301, vol. 21.
Shiji 史記. Compiled by Sima Tan 司馬談 and Sima Qian 司馬遷. Beijing: Zhonghua shuju, 1959.
Shi ji zhuan 詩集傳. By Zhang Zai 張載. Taipei: Taiwan shangwu yinshuguan, 1981.
Shi ji zhuan 詩集傳. By Zhu Xi 朱熹. Compiled and Annotated by Wang Yi 王逸 and Hong Xingzu 洪興祖. Changsha: Yuelu shushe, 1989.
Sifen lü 四分律 (Skt. Dharmagupta-Vinya). Translated by Buddhayaśas 佛陀耶舍 and Fonian 竺佛念. T. 22, no. 1428.
Sifen lü biqiu hanzhu jieben 四分律比丘含注戒本. By Daoxuan 道宣. T. 40, no. 1806.
Songshu 宋書. Compiled by Shen Yue 沈約. Beijing: Zhonghua shuju, 1974.
Sui shu 隋書. By Fang Xuanling 房玄齡 et al. Beijing: Zhonghua shuju, 1973.
Taiping guanji 太平廣記. Edited by Li Fang 李昉 et al. Shanghai: Shanghai guji chubanshe, 1990.
Taiping yulan 太平御覽. Edited by Li Fang 李昉 et al. Beijing: Zhonghua shuju, 1960.
Tiandi ruixiang zhi 天地瑞祥志. Compiled by Sa Shouzhen 薩守鎮 (fl. 666 CE). In *Xijian Tangdai tianwen shiliao sanzhong* 稀見唐代天文史料三種. Edited by Gao Keli 高柯立. Beijing: Guojia tushuguan chubanshe, 2011, 1–584.
Tiantai sijiaoyi 天臺四要儀. By Diguan 諦觀. T. 46, no. 1931.
Tō daiwajō tosei den 唐大和尚東征傳. By Genkai 元開. T. no. 2089, vol. 51.
Weilüe 魏略. Compiled by Yu Huan 魚豢. Taipei: Dingwen shuju, 1990.
Weimo yi ji 維摩義記. By Huiyuan 慧遠. T. 38, no. 1776.
Wenxian tongkao 文獻通考. Compiled by Ma Duanlin 馬端臨 (1254–1324 CE). Shanghai: Shanghai guji chubanshe, 1987.
Wuxing dayi 五行大義. By Xiao Ji 蕭吉. Edited by Qian Hang 錢杭. Shanghai shudian chubanshe, 2001.
Xianguo suilu 現果隨錄. By Jiexian 戒顯. X. 88, no. 1642.
Xiangyu shu 相雨書. By Huang Zifa 黃子發. Shanghai: Shangwu yinshu guan, 1939.
Xin Tangshu 新唐書. Compiled by Ouyang Xiu 歐陽修, Song Qi 宋祁, et al. Beijing: Zhonghua shuju, 1975.
Xiuxi zhiguan zuochan fayao 修習止觀坐禪法要. By Zhiyi 智顗. T. 46, no. 1915.
Xuanguai lu 玄怪錄. By Niu, Sengru 牛僧孺. Shanghai: Shanghai guji chubanshe, 1985.
Yixue jicheng 易學集成. By Yang Shiwen 杨世文, et al. Chengdu: Sichuan daxue chubanshe, 1998.
Yue jue shu 越絕書. By Wu Ping 吳平. Suppelmented by Yuan Kang 袁康. Annotated by Li Bujia 李步嘉. Beijing: Zhonghua shuju, 2018.
Yuhan shanfang ji yishu 玉函山房輯佚書. Compiled by Ma Guohan 馬國翰 (1794–1857). 1874; rpt. Taipei: Wenhai chubanshe, 1967.
Yulizi 郁離子. By Liu Ji 劉基. Annotated by Wei Jianxian 魏建猷 and Xiao Jixiang 蕭善鄉. Shanghai: Shanghai guji chubanshe, 1981.
Yunji qiqian 雲笈七籤. Compiled by Zhang Junfang 張君房, in Zhengtong daozang 正統道藏, vol. 37. Taipei: Xinwenfeng chuban gongsi, 1985.
Zhiguan fuxing chuanhong jue 止觀輔行傳弘訣. By Zhanran 湛然. T. 46, no. 1912.
Zhuangzi jishi 庄子集释. By Guo Qingfan 郭庆藩. Annotated by Wang Xiaoyu 王孝雨. Beijing: Zhonghua shuju, 1961.
Zuben Shanhaijing Tuzan 山海經圖讚. By Guo Pu 郭璞. Annotated by Zhang Zongxiang 張宗祥. Shanghai: Gudian wenxian chubanshe, 1958.

Secondary Sources

Ahern, Emily M. *The Cult of the Dead in a Chinese Village.* Stanford: Stanford University Press, 1973.

Allon, Mark. *Three Gāndhārī Ekottarikāgama-Type Sūtras: British Library Kharoṣṭhī Fragments 12 and 14*, Gandhāran Buddhist Texts vol. 2. Seattle: University of Washington Press, 2001.

Barendrgt, Bart. "A Supernatural Topography of the Southern Sumatran Highlands." *Crossroads: An Interdisciplinary Journal of Southeast Asian Studies* 18, no. 1 (2006): 113–25.

Barker, Randolph. "The Origin and Spread of Early-Ripening Champa Rice: Its Impact on Song Dynasty China." *Rice* 4 (2011): 184–86.

Barrett, Timothy H. *The Women Who Discovered Printing.* New Haven: Yale University Press, 2008.

——— "Woodblock Dyeing and Printing Technology in China, c. 700 A.D.: The Innovations of Ms. Liu, and Other Evidence." *Bulletin of the School of Oriental and African Studies* 64, no. 2 (2001): 240–47.

Bartlett, Robert. *The Natural and the Supernatural in the Middle Ages.* The Wiles Lectures given at the Queen's University of Belfast, 2006. Cambridge: Cambridge University Press, 2008.

Beal, Samuel trans. *Xi you ji: Buddhist Records of the Western World.* Volume 1. London: Kegan Paul, Trench, Trübner & Co., 1906.

Benn, James A. *Tea in China: A Religious and Cultural History.* Honolulu: University of Hawai'i Press, 2015.

Bhikkhu Ñāṇamoli and Bhikkhu Bodhi trans. *The Middle Length Discourses of the Buddha: A Translation of the Majjhima Nikaya.* Boston: Wisdom Publications, 1995.

Bian, Lianquan 邊連全 ed. *Zhu siyang jishu* 豬飼養技術 (*Pig Feeding Technology*). Shenyang: Dongbei daxue chubanshe, 2010.

Bildhauer, Bettina. *The Monstrous Middle Ages.* Toronto: University of Toronto Press, 2004.

Bokenkamp, Stephen R. "The Herb Calamus and the Transcendent Han Zhong in Taoist Literature." *Studies in Chinese Religions* 1, no. 4 (2015): 293–305.

Boyce, Mary. *A Reader in Manichean Middle Persian and Parthian.* Leiden: Brill, 1975.

Boyd, James W. *Satan and Mara: Christian and Buddhist Symbols of Evil.* Leiden: Brill, 1975.

Bracke, Marc B. M. "Review of Wallowing in Pigs: Description of the Behaviour and Its Motivational Basis." *Applied Animal Behaviour Science* 132, nos. 1–2 (2010): 1–13.

Bray, Francesca. *Technology and Gender: Fabrics of Power in Late Imperial China.* Berkeley: University of California Press, 1997.

——— "Essence and Utility: The Classification of Crop Plants in China." *Chinese Science* 9 (1989): 1–13.

——— *Science and Civilisation in China. Volume 6: Biology and Biological Technology. Part II: Agriculture.* Cambridge: Cambridge University Press, 1984.

Brighenti, Francesco. "Kradi Mliva: The Phenomenon of Tiger-Transformation in the Traditional Lore of the Kondh Tribes of Orissa." *Lokaratna* 4 (2011): 11–25.

Cai, Hongsheng 蔡鴻生. *Tangdai jiuxinghu yu Tujue wenhua* 唐代九姓胡與突厥文化 (*The Nine-Surname Sogdians and Turkic Culture in the Tang Dynasty*). Beijing: Zhonghua shuju, 1998.

Cai, Zhemao 蔡哲茂, "Jiaguwen sifang feng ming zaitan 甲骨文四方風名再探 (Revisiting the Names of Four-Directional Winds in the Oracle Bone Inscriptions)." In *Jin Xiangheng jiaoshou shishi zhounian jinian lunwenji* 金祥恆教授逝世週年紀念論文集 (Collected Papers in memory of the one-year anniversary of the Passing of Professor Jin Xiangheng). 123–52. Taipei: Jin Xiangheng jiaoshou shishi zhounian jinian lunwenji bianji xiaozu, 1990.

Cammiade, L. A. "Man-Eaters and Were-Tigers." *Man* 31 (1931): 217–20.

Cao, Deborah. *Animals in China: Law and Society*. Basingstoke: Palgrave Macmillan, 2015.

Capper, Daniel. *Roaming Free Like a Deer: Buddhism and the Natural World*. Ithaca: Cornell University Press, 2022.

Chavannes, Édouard and Paul Pelliot. "Un traité manichéen retrouvé en Chine I." *Journal Asiatique* 10, no. 18 (1911): 499–617.

Chee, Liz P. Y. *Mao's Bestiary: Medicinal Animals and Modern China*. Durham, NC: Duke University Press, 2021.

Chen, Huaiyu 陳懷宇. *In the Land of Tigers and Snakes: Living with Animals in Medieval Chinese Religions*. New York: Columbia University Press, 2023.

——— "Zoomancy/divination by animals." In *Handbook of Chinese Divination Techniques*, edited by Stephen N. Kory. Leiden: Brill, forthcoming.

——— Review. Yan Liu, *Healing with Poisons: Potent Medicines in Medieval China*. *China Review International* 27, no. 2 (2020): 124–27.

——— Review. *Mao's Bestiary: Medicinal Animals and Modern China*. By Liz P. Y. Chee. In *Chinese Historical Review* 29, no. 1 (2022): 67–69.

——— "Plant Science and Technology in Medieval China." In *A Cultural History of Plants in the Post-Classical Era*, edited by Alain Touwaide, 79–99. London: Bloomsbury, 2022.

——— "The Changing Images of Zodiac Animals in Medieval Chinese Buddhist Literature." *Nesir Journal of Literary Studies* 1 (2021): 37–54.

——— "The Other as the Transformed Alliance: Living with the Tiger in Medieval Chinese Daoism." *Polylog. Zeitschrift für interkulturelles Philosophieren* 45 (2021): 4–22.

——— *Dongwu yu zhonggu zhengzhi zongjiao zhixu* 動物與中古政治宗教秩序 (*Animals and Political and Religious Order in Medieval China*). Shanghai: Shanghai guji chubanshe, 2012, expanded edition, 2021.

——— Review. *Animals through Chinese History: Earliest Times to 1911*. Eds. Roel Sterckx, Martina Siebert, and Dagmar Schäfer. In *Journal of the Royal Asiatic Society*, 3rd series, 30, no. 2 (2020): 376–78.

——— Review. *Insects in Chinese Literature: A Study and Anthology*. By Wilt L. Idema. In *China Review International* 26, nos. 1 & 2 (2019): 78–81.

——— Review. *Panda Nation: The Construction and Conservation of China's Modern Icon*. By E. Elena Songster. In *China Review International* 26, nos. 1 & 2 (2019): 110–14.

——— "A Buddhist Classification of Plants and Animals in Early Tang China." *Journal of Asian History* 43, no. 1 (2009): 31–51.

——— "Daoxuan yu Sun Simiao yixue jiaoliu zhi yizheng lice 道宣與孫思邈醫學交流之一證蠡測 (A Preliminary Study on a Case of the Exchange of Medical Knowledge between Daoxuan and Sun Simiao)." *Dunhuang Tulufan yanjiu* 敦煌吐魯番研究 9 (2006): 403–8.

Chen, Jinhua 陳金華. "Dongya fojiao zhong de biandi qingjie: lun shengdi ji zupu de jiangou 東亞佛教中的邊地情結：論聖地及祖譜的建構 (The "Borderland Complex" in East Asian Buddhism: A Discussion of the Development of Sacred Space and Genealogies)." *Foxue yanjiu* 佛學研究 21 (2012): 22–41.

——— "One Name, Three Monks: Two Northern Chan Masters Emerge from the Shadow of Their Contemporary, the Tiantai Master Zhanran (711–782)." *Journal of International Association for Buddhist Studies* 22, no. 1 (1999): 1–91.

Chen, Sanping 陳三平. *Multicultural China in the Early Middle Ages*. Philadelphia: University of Pennsylvania Press, 2012.

Chen, Weiji 陳維稷. 1984. *Zhongguo fangzhi kexue jishu shi* 中國紡織科學技術史 (*A History of Textile Production Technology in China*). Beijing: Kexue chubanshe.

Chen, Yiyuan 陳益源 and Xu Jincheng 徐錦成. "Lun yuwai hanwen biji xiaozhong de hu 論域外漢文筆記小說中的虎 (A Study on the Tiger in the Chinese Tales in Foreign Regions)." In *2007 Dongya hanwenxue yu minus wenhua guoji xueshu yantaohui lunwenji* 2007 東亞漢文學與民俗文化國際學術研討會論文集 (Collected Papers from the International Conference on East Asian Chinese Texts and Folklore Culture). Eds. Wang Sanqing and Chen Yiyuan, 425–41. Taipei: Lexue shuju, 2007.

Cheng, Guofu 程國賦. *Tang Wudai xiaoshuo de wenhua chanshi* 唐五代小說的文化闡釋 (*The Cultural Interpretation on the Novels in the Tang and Five Dynasties*). Beijing: Renmin wenxue chubanshe, 2002.

Chi, Limei (Chi Reibai) 池麗梅. *Tōdai tendai bukkyō fukkō undō kenkyū josetsu: Keikei tannen to sono shikan fukōden guketsu* 唐代天台佛教復興運動研究序說：荊渓湛然とその止観輔行伝弘決 (A Preliminary Study on the Revival Movement of the Tiantai Buddhism in the Tang Dynasty: Jingxi Zhanran and His *Zhiguan fuxing chuan hongjue*). Tokyo: Daizō Shuppan, 2008.

Chu, Zexiang 儲澤祥. "Hanyu kongjian fangwei duanyu de lishi yanbian de jige tedian 漢語空間方位短語的歷史演變的幾個特點 (Several Features on the Historical Development of the Phases on Cardinal Directions and Spaces in Chinese Language)." *Gu hanyu yanjiu* 古漢語研究 30, no. 1 (1996): 57–61.

Clark, Robert Warren. "Mara: Psychopathology and Evil in the Buddhism of India and Tibet." PhD dissertation, University of Virginia, 1994.

Cohen, Jeffrey Jerome. *Of Giants: Sex, Monsters, and The Middle Ages*. Minneapolis: University of Minnesota Press, 1999.

Culver, Annika A. *Japan's Empire of Birds: Aristocrats, Anglo-Americans, and Transwar Ornithology*. London: Bloomsbury, 2022.

Dale, Stephen F. "Silk Road, Cotton Road or… Indo-Chinese Trade in Pre-European Times." *Modern Asian Studies* 43, no. 1 (2009): 79–88.

Dalton, Jacob and Sam van Schaik. *Tibetan Tantric Manuscripts from Dunhuang*. Leiden: Brill, 2006.

de Groot, Jan J. *The Religious System of China*. Volume 4. Leiden: Brill, 1901.

Demarée, Gaston R. & This Rutishauser. "From 'Periodical Observations' to 'Anthochronology' and 'Phenology': The Scientific Debate between Adolphe Quetelet and Charles Morren on the Origin of the Word 'Phenology'." *International Journal of Biometeorology* 55, no. 6 (2011): 753–61.

de Crespigny, Rafe Richard trans. *Emperor Huan and Emperor Ling: Being the Chronicle of Later Han for the Years 157 to 189 AD as Recorded in Chapters 54 to 59 of the Zizhi Tongjian of Sima Guang*. Canberra: Australian National University, 1989.

——— *Portents of Protest in the Later Han Dynasty: The Memorials of Hsiang K'ai to Emperor Huan*. Canberra: Australia National University Press, 1976.

De Pee, Christian. "Wards of Words: Textual Geographies and Urban Space in Song-Dynasty Luoyang, 960–1127." *Journal of the Economic and Social History of the Orient* 52, no. 1 (2009): 85–116.

Diener, Paul et al. "Ecology, Evolution, and the Search for Cultural Origins: The Question of Islamic Pig Prohibition [and Comments and Reply]." *Current Anthropology* 19, no. 3 (1978): 493–540.

Diény, Jean-Pierre. *Le symbolisme du dragon dans la Chine antique*. Paris: Collège de France, Institut des hautes études chinoises, 1987.

Dong, Naibin 董乃斌. *Zhongguo gudian xiaoshuo de wenti duli* 中國古典小說的文體獨立 (*The Classical Novels as Independence Genre in China*). Beijing: Zhongguo shehui kexue chubanshe, 1994.

Drège, Jean-Pierre. "L'analyse fibreuse des papiers et la datation des manuscrits de Dunhuang." *Journal Asiatique* 274 (1986): 403–15.

Duan, Qing 段晴 and Zhang Zhiqing 張志清 eds. *Zhongguo guojia tushuguan cang xiyu wenshu: Fanwen, quluwen juan* 中國國家圖書館藏西域文書：梵文、佉盧文卷 (The Manuscripts from the Western Regions Preserved at the National Library of China: The Sanskrit and Kharoṣṭhī Volume). Shanghai: Zhongxi shuju, 2013.

Dull, Jack L. "A Historical Introduction to the Apocryphal (Ch'an-Wei) Texts of the Han Dynasty." PhD dissertation, University of Washington, 1966.

Dutt, Nalinaksha. *Gilgit Manuscripts.* Delhi: Sri Satguru Publications, Delhi 1984, reprinted.

Egan, Ronald. *The Problem of Beauty: Aesthetic Thought and Pursuits in Northern Song Dynasty China.* Cambridge, MA: Harvard University Asia Center, 2006.

Elvin, Mark. *The Retreat of the Elephants: An Environmental History of China.* New Haven: Yale University Press, 2004.

Engelhardt, Ute. "Dietetics in Tang China and the First Extant Works of Materia Dieterica." In *Innovation in Chinese Medicine*, edited by Elisabeth Hsu, 173–91. Cambridge: Cambridge University Press, 2001.

Essig, Mark. *Lesser Beasts: A Snout-to-Tail History of the Humble Pig.* New York: Basic Books, 2015.

Feng, Hanyong 馮漢鏞. "Tang Song shidai de zaochuanye 唐宋時代的造船業 (Shipbuilding in the Tang and Song Dynasties)." *Lishi jiaoxue* 歷史教學 (*Teaching History*) no. 10 (1957): 10–14.

Feng, Jiren. 2007. "Bracketing Likened to Flowers, Branches and Foliage: Architectural Metaphors and Conceptualization in Tenth to Twelfth-Century China as Reflected in the *Yingzao Fashi*." *T'oung Pao* 2nd Series, 93, no. 4/5 (2007): 369–432.

Feng, Guangping 馮廣平. "Tangdai Pingquan shanju zaipei zhiwu kao 唐代平泉山居栽培植物考 (Textual Research on Cultivated Plants in Pingquan Villa of Tang Dynasty)." *Zhongguo lishi dili luncong* 中國歷史地理論叢 (*Journal of Chinese Historical Geography*) 31, no. 1 (2016): 60–72.

Forte, Antonino. *Mingtang and Buddhist Utopias in the History of the Astronomical Clock: The Tower, Statue, and Armillary Sphere Constructed by Empress Wu.* Roma: Istituto Italiano per il Medio ed Estremo Oriente, and Paris: École française d'Extrême-Orient, 1988.

Freedman, Maurice. "Ancestor Worship: Two Facets of the Chinese Case." In *Social Organization: Essays presented to Raymond Firth*, edited by Maurice Freedman, 85–103. Chicago: Aldine Publishing, 1967.

Friedman, John B. *The Monstrous Races in Medieval Art and Thought.* Syracuse: Syracuse University Press, 2nd edition, 2000.

Fudge, Erica. "A Left-Handed Blow: Writing the History of Animals." In *Representing Animals*, edited by Nigel Rothfels, 3–18. Bloomington: Indiana University Press, 2002.

Fujiansheng Quanzhou haiwai jiaotongshi bowuguan 福建省泉州海外交通史博物館 ed. *Quanzhouwan Songdai haichuan fajue yu yanjiu* 泉州灣宋代海船發掘與研究 (*The Discovery and Study on the Sea Boat in the Song Dynasty from the Quanzhou Bay*). Beijing: Haiyang chubanshe, 2017.

Gan, Bao 干寶. *In Search of the Supernatural: The Written Record.* Translated by Kenneth J. DeWorskin and J. I. Crump, Jr. Stanford: Stanford University Press, 1996.

Gao, Keli 高柯立 ed. *Xijian Tangdai tianwen shiliao sanzhong* 稀見唐代天文史料三種 (*Three Rare Historical Sources on Astronomy in the Tang Dynasty*). Beijing: Guojia tushuguan chubanshe, 2011.

Ge, Quansheng 葛全勝 et al. "Sui Tang shqi dongzhongbu diqu wendu bianhua de chongjian 隋唐時期東中部地區溫度變化的重建 (The Reconstruction of the Changing Temperatures in the Eastern and Central Regions of the Sui and Tang Periods)." *Kexue tongbao* 科學通報 (*Science Bulletin*) 55, no. 31 (2010): 3048–52.

Ge, Zhaoguang 葛兆光. *Zhaizi zhongguo: Chongjian youguan zhongguo de lishi lunshu* 宅茲中國: 重建有關中國的歷史論述 (*Here in China I Dwell: Reconstructing Historical Discourses of China for Our Time*). Beijing: Zhonghua shuju, 2011.

Gilmore, David D. *Monsters: Evil Beings, Mythical Beasts, and All Manner of Imaginary Terrors*. Philadelphia: University of Pennsylvania Press, 2009.

Golas. Peter J. "Rural China in the Song." *Journal of Asian Studies* 39, no. 2 (1980): 291–325.

Graham, A. C. *Yin-Yang and the Nature of Correlative Thinking*. Singapore: Institute of East Asian Philosophies, 1986.

Griffiths, Paul J. trans. *The Realm of Awakening: A Translation and Study of the Tenth Chapter of Asaṅga's Mahāyānasaṅgraha*. Oxford: Oxford University Press, 1989.

Gross, Aaron. *The Question of the Animal and Religion: Theoretical Stakes, Practical Implications*. New York: Columbia University Press, 2014.

Guo, Qinghua. "*Yingzao Fashi:* Twelfth-Century Chinese Building Manual." *Architectural History* 41 (1998): 1–13.

Hammond, Charles E. "Excursions in Tiger Lore." *Asia Major* 4, no. 1 (1991): 87–100.

——— "Sacred Metamorphosis: The Weretiger and the Shaman." *Acta Orientalia Hungaricae* 46, nos. 2–3 (1992–1993): 235–55.

——— "The Demonization of the Other: Women and Minorities as Weretigers." *Journal of Chinese Religions* 23, no. 1 (1995): 59–80.

Haraway, Donna J. *When Species Meet*. Minneapolis and London: University of Minnesota Press, 2008.

Hargett, James M. trans. *Treatises of the Supervisor and Guardian of the Cinnamon Sea*. By Fan Chengda. Seattle: University of Washington Press, 2010.

——— "Playing Second Fiddle: The Luan-Bird in Early and Medieval Chinese Literature." *T'oung Pao*, 2nd series, 75, nos. 4–5 (1989): 235–62.

Harper, Donald. "The Textual Form of Knowledge: Occult Miscellanies in Ancient and Medieval Chinese Manuscripts, 4th Century BCE to 10th Century CE." In *Looking at It from Asia: The Processes that Shaped the Sources of History of Science*, edited by Florence Bretelle-Establet, 37–80. Dordrecht: Springer, 2010.

——— "Flowers in T'ang Poetry: Pomegranate, Sea Pomegranate, and Mountain Pomegranate." *Journal of the American Oriental Society* 106, no. 1 (1986): 139–53.

Harris, Marvin. *Cows, Pigs, Wars, and Witches: The Riddles of Culture*. New York: Random House, 1974.

Hardy, Anne. "Animals, Disease and Man: Making Connections." *Perspectives in Biology and Medicine* 46 (2003): 200–215.

Hartmann, Jens-Uwe and Chanwit Tudkeao. "Three Sanskrit Fragments of the Ratnaketuparivarta." In *The British Library Sanskrit Fragments*, vol. 2, edited by Seishi Karashima and Klaus Wille, 589–96. Tokyo: International Research Institute for Advanced Buddhology, Soka University, 2009.

Hartwell, Robert M. "Demographic, Political, and Social Transformations of China, 750–1550." *Harvard Journal of Asiatic Studies* 42, no. 2 (1982): 365–442.

Haw, Stephen G. "Tung Oil and tong Trees." *Zeitschrift der Deutschen Morgenländischen Gesellschaft* 167, no. 1 (2017): 215–36.

He, Fangyao 何方耀. "Han Tang Zhongguo fomen de biandi yishi yu fanyu xuexi rechao 漢唐中國佛門的邊地意識與梵語學習熱潮." *Jiuzhou xuelin* 九州學林 3, no. 4 (2005): 137–40.

Helman-Ważny, Agnieszka. "More than Meets the Eye: Fibre and Paper Analysis of the Chinese Manuscripts from the Silk Roads." *STAR: Science & Technology of Archaeological Research* 2, no. 2 (2016): 127–40.

Hesse, Brian. "Pig Lovers and Pig Haters: Patterns of Palestinian Port Production," *Journal of Ethnobiology* 10, no. 2 (1990): 195–225.

Hishikawa, Akiko 菱川晶子. *Ōkami no minzokugaku: jinjū kōshōshi no kenkyū* 狼の民俗学: 人獣交渉史の研究 (*The Folklore of Wolf: A Study on the Interactions between Humans and Beasts*). Tokyo: Tōkyōdaigaku shuppankai, 2018.

Ho, Peng Yoke 何丙郁. *Chinese Mathematical Astrology: Reaching Out to the Stars*. London: Routledge, 2004.

——— *Li, Qi and Shu: An Introduction to Science and Civilization in China*. Hong Kong: Hong Kong University Press, 1985.

Hoernle, A. F. Rudolf. *Manuscript Remains of Buddhist Literature found in Eastern Turkestan*. Oxford: Clarendon Press, 1916.

Holden, Palmer J. and M. E. Ensminger. *Swine Science*. Princeton: Prentice Hall, 7th edition, 2005.

Hong, Ruiying 洪瑞英. *Zhongguo renhu bianxing gushi yanjiu* 中國人虎變形故事研究 (A Study on the Tales on the Transformations between Humans and Tigers in China). Xinbei: Huamulan wenhua chubanshe, 2011.

Horace. *The Works of Horace*. Epistle II. Translated by C. Smart, revised by Theodore Alois Buckley. London: George Bell & Sons, 1888.

Hou, Ching-lang 侯錦郎. "The Chinese Belief in Baleful Stars." In *Facets of Taoism: Essays in Chinese Religion*, edited by Holmes Welch and Anna Seidel, 209–19. New Haven and London: Yale University Press, 1979.

Hou, Yongjian 侯甬堅, Cao Zhihong 曹志紅, Zhang Jie 張潔, and Li Ji 李冀. *Zhongguo huanjingshi yanjiu* 中國環境史研究 (*Studies on Chinese Environmental History*), vol. 3: Lishi dongwu yanjiu zhuanji 歷史動物研究專輯 (Special Volume on Historical Animals). Beijing: Zhongguo huanjing kexue chubanshe, 2014.

Hu, Shiu-Ying. "The Economic Botany of the Paulownias." *Economic Botany* 15, no. 1 (1961): 11–27.

Huang, Chunyan 黃純艷. *Zaochuanye shiyu xia de Songdai shehui* 造船業視域下的宋代社會 (*Society in Song China: A Perspective of Shipbuilding*). Shanghai: Shanghai renmin chubanshe, 2017.

Huang, Yong 黃勇. "Cheng Yi's 程頤 Moral Philosophy." In *Dao Companion to Neo-Confucian Philosophy*, edited by John Makeham, 59–87. Dordrecht: Springer, 2010.

Huang, Yongwu 黃永武 ed. *Dunhuang baozang* 敦煌寶藏 (*Treasures from Dunhuang*). Taipei: Xinwenfeng chuban gongsi, 1986.

Huang Zhengjian 黃正建. *Dunhuang zhanbu wenshu yu Tang Wudai zhanbu yanjiu* 敦煌占卜文書與唐五代占卜研究 (A Study on the Divination Manuscripts from Dunhuang and the Divination Culture in the Tang and Five Dynasties). Beijing: Zhongguo shehui kexue chubanshe, 2014.

Huang, Zhifan 黃志繁. "Shanshou zhijun, huhuan yu daode jiaohua: cezhong yu Ming Qing nanfang shanqu 山獸之君、虎患與道德教化：側重於明清南方地區 (The Lord of Mountain Beasts, Tiger Disaster, and Moral Transformation: With Special Reference to Southern Regions in the Ming and Qing Dynasties)." *Zhongguo shehui lishi pinglun* 中國社會歷史評論 (*Chinese Social Historical Review*) 7 (2006): 1–16.

Hutton, J. H. "Leopard-Men in the Naga Hills." *The Journal of the Royal Anthropological Institute of Great Britain and Ireland* 50 (1920): 41–51.

Hwa, Yol Jung. "Edouard Glissant's Aethetics of Relation as Diversity and Creolization." In *Postcolonialism and Political Theory*, edited by Nalini Therese Persram, 193–225. Lanham, MD: Lexington Books, 2007.

Idema, Wilt L. *Insects in Chinese Literature: A Study and Anthology*. Cambria Sinophone World Series. Amherst, New York: Cambria, 2019.

Ienaga, Masaki 家永真幸. *Kokuhō no seijishi: "Chūgoku" no Kokyū to panda* 国宝の政治史: "中国"の故宮とパンダ (*The Political History of Natural Treasures: China's Palace Museum and Panda*). Tokyo: Tōkyō Daigaku Shuppankai, 2017.

——— *Panda gaikō* パンダ外交 (Panda Diplomacy). Tōkyō: Mediafakutorī, 2011.

Ikram, Salima. *Choice Cuts: Meat Production in Ancient Egypt*. Leuven: Peeters, 1995.

Ingram, D. L. "Evaporative cooling in pig." *Nature* 207 (1965): 415–16.

Jensen. Lionel M. "Legends of Confucius." In *Hawaii Reader in Traditional Chinese Culture*, edited by Victor H. Mair, Nancy S. Steinhardt, and Paul R. Goldin, 234–38. Honolulu: University of Hawai'i Press, 2005.

——— "The Genesis of Kongzi in Ancient Narrative: The Figurative as Historical." In *On Sacred Grounds: Culture, Society, Politics, and the Formation of the Cult of Confucius*, edited by Thomas A. Wilson, 175–221. Cambridge, MA: Harvard University Press, 2002.

Jin, Chengyu 金程宇. *Yuwai hanji congkao* 域外漢籍叢考 (*Studies on the Chinese Texts in Foreign Regions*). Beijing: Zhonghua shuju, 2007.

Jones, Timothy S. and David A. Sprunger eds. *Marvels, Monsters, and Miracles: Studies in the Medieval and Early Modern Imaginations*. Kalamazoo: Medieval Institute Publications, 2002.

Kaderas, Christoph. *Die Leishu der imperialen Bibliothek des Kaisers Qianlong (reg. 1736–1796): Untersuchungen zur chinesischen Enzyklopädie*. Wiesbaden: Otto Harrassowitz Verlag, 1998.

Kakuitsu, Shoji 荘司格一. "Jintoraden seiritsu no haikei chusei ni okeru tora no setsu 人虎伝」成立の背景--中世における虎の説話 (The Background of the Rise of the Weretigers: Tales on the Tiger in the Middle Ages)." *Yamagata daigaku kiyo* 山形大学紀要 9, no. 4 (1981): 417–43.

Kalinowski, Marc, ed. *Divination et société dans la Chine médiévale: Étude des manuscrits de Dunhuang de la Bibliothèque nationale de France et de la British*. Paris: Bibliothèque nationale de France, 2003.

Kang, Xiaofei 康笑菲. *The Cult of the Fox: Power, Gender, and Popular Religion in Late Imperial and Modern China*. New York: Columbia University Press, 2006.

Karashima, Seishi 辛嶋靜志. "Who Composed the Mahāyāna Scriptures?—The Mahāsāṃghikas and Vaitulya Scriptures*," *Annual Report of the International Research Institute for Advanced Buddhology at Soka University for the Academic Year 2014* 18 (2015): 113–62.

——— "The Sanskrit Fragments Or.15010 in the Hoernle Collection." In *The British Library Sanskrit Fragments*, vol. 2, edited by Seishi Karashima and Klaus Wille, 335–550. Tokyo: International Research Institute for Advanced Buddhology, Soka University, 2009.

——— "Four Sanskrit Fragments of the Ratnaketuparivarta in the Stein Collection." In *The British Library Sanskrit Fragment: Buddhist Manuscripts from Central Asia*, vol. 1, edited by Seishi Karashima and Klaus Wille, 177–89. Tokyo: The International Research Institute for Advanced Buddhology, Soka University, 2006.

Kean, Hilda. "Animal-Human Histories." In *New Directions in Social and Cultural History*, edited by Sasha Handley, Rohan McWilliam, and Lucy Noakes, 173–89. London: Bloomsbury, 2018.

Kern, Martin. "Early Chinese Divination and Its Rhetoric." In *Coping with the Future: Theories and Practices of Divination in East Asia*, edited by Michael Lackner, 255–88. Leiden: Brill, 2017.

Kharmawphlang, Desmond. "In Search of Tigermen: The Were-Tiger Tradition of the Khasis." *India International Centre Quarterly* 27, no. 4/ 28 , no. 1 (2000–2001): 160–76.

Kirk, Robert G. W. and Michael Worboys. "Medicine and Species: One Medicine, One History?" In *The Oxford Handbook of the History of Medicine*, edited by Mark Jackson, 561–77. Oxford: Oxford University Press, 2011.

Klimkeit, Hans-Joachim translated and presented. *Gnosis on the Silk Road: Gnostic Texts from Central Asia.* San Francisco: HarperSanFrancisco, 1993.

Knapp, Keith. "The Use and Understanding of Domestic Animals in Medieval Northern China." *Early Medieval China* 25 (2019): 85–99.

Knechtges, David R., Taiping Chang eds., *Ancient and Early Medieval Chinese Literature (Vol. I): A Reference Guide*, Part One. Leiden: Brill, 2010.

Kohn, Euardo. *How Forests Think: Toward an Anthropology beyond the Human.* Berkeley: University of California Press, 2013.

Kōno Kimiko 河野喜美子. *Nihon ryōiki to Chūgoku no denshō* 日本霊異記と中国の伝承 (*The Record of Miraculous Stories* in Japan and the Chinese Transmissions). Tōkyō: Benseisha, 1996.

Kuhn, Dieter. *Science and Civilisation in China.* Volume 5. Chemistry and Chemical Technology. Part 9. Textile Technology: Spinning and Reeling. Cambridge: Cambridge University Press, 1988.

Kurumiya, Yenshu 久留宮圓秀. *Ratnaketuparivarta: Sanskrit Text.* Kyoto: Heirakuji Shoten, 1978.

Li, Fan 李璠. *Zhongguo zaipei zhiwu fazhan shi* 中國栽培植物發展史 (*A History of the Development of Cultivated Plants in China*). Beijing: Kexue chubanshe, 1984.

Li, Fengmao 李豐楙. "Liuchao jingguai dongwu yu Daojiao fashu sixiang 六朝精怪動物與道教法術思想 (The Demonic Animals and the Daoist Rituals and Thoughts in the Six Dynasties)." *Zhongguo gudian xiaoshuo zhuanji* 中國古典小說專輯 3 (1981): 1–36.

Li, Hui-lin. "Hallucinogenic Plants in Chinese Herbals." *Botanical Museum Leaflets* (Harvard University) 25, no. 6 (1977): 161–81.

——— "The Origin and Use of Cannabis in Eastern Asia Linguistic-Cultural Implications." *Economic Botany* 28, no. 3 (1974): 293–301.

——— "An Archaeological and Historical Account of Cannabis in China." *Economic Botany* 28, no. 4 (1974): 437–48.

——— "The Vegetables of Ancient China." *Economic Botany* 23, no. 3 (1969): 253–60.

Li, Qiwen 李啟文, "Lun guren shi ruhe kandai fangwei de 論古人是如何看待方位的 (How Did Ancient People View the Cardinal Directions)." *Xueshu yanjiu* 學術研究 11, no. 2 (1999): 72–75.

Lin, Fushi 林富士. "Renjian zhi mei: Han Tang zhiji de jingmei gushi xilun 人間之魅：漢唐之際的精魅故事析論 (The Enchantment within the Human Realm: A Analytical Discussion on the Tales about the Spirits of Enchantment between the Han and Tang Periods)." *Zhongyanyuan shiyusuo jikan* 中研院史語所集刊 78, no. 1 (2007): 107–82.

Lin, Liming 林禮明. *Guiyu shijie* 鬼魊世界 (*The World of Ghosts and Demons*). Xiamen: Xiamen daxue chubanshe, 1993.

Lin, Lisheng 林立勝 and Ye Chaolong 葉超龍. "Hebo shenhua youlai ji xingxiang chayi tantao 河伯神話由來及形象差異探討 (A Study on the Origin of the Duke of River and Its Various Images)." *Shenhua yu wenxue lunwen xuanji 2012–2013* 神話與文學論文選輯 2012–2013 (*Selected Papers on Myths and Literature*), edited by Liu Yanping

劉燕萍, 82–89. Hong Kong: Lingnan University, Department of Chinese Literature, 2012–2013.

Lin, Wushu 林悟殊, Monijiao jiqi dongjian 摩尼教及其東漸 (*Manichaeism and Its Spread to the East*). Beijing: Zhonghua shuju, 1987.

Ling, Travor Oswald. *Buddhism and the Mythology of Evil: A Study in Theravada Buddhism*. London: Allen & Unwin, 1962.

Lionarons, Joyce Tally. "From Monster to Martyr: The Old English Legend of Saint Christopher." In *Marvels, Monsters, and Miracles: Studies in the Medieval and Early Modern Imaginations*, edited by Timothy S. Jones and David A. Sprunger, 167–82. Kalamazoo: Medieval Institute, 2002.

Lippiello, Tiziana. *Auspicious Omens and Miracles in Ancient China: Han, Three Kingdoms and Six Dynasties*. Sankt Augustin: Monumenta Serica Institut, 2001.

Liu, Changzhi. 1994. "Zhongguo chuantong zhiwuxue de fazhan 中國傳統植物學的發展 (The Development of Chinese Traditional Plant Science)." In *Zhongguo zhiwu xue shi* 中國植物學史 (*A History of Plant Science in China*), edited by Zhongguo zhiwuxue xuehui (Chinese Association for Plant Science), 35–59. Beijing: Kexue chubanshe, 1994.

Liu, Fei-hu, Xia Chen, Bo Lng, Rui-yan Shuai, and Chun-lin Long. "Historical and Botanical Evidence of Distribution, Cultivation and Utilization of *Linum usitatissimum* L. (flax) in China." *Vegetation History and Archaeobotany* 20, no. 6 (2011): 561–66.

Liu, Fei-hu, Hua-ren Hu, Gang-hui Du, Gang Deng and Yang Yang. "Ethnobotanical Research on Origin, Cultivation, Distribution and Utilization of Hemp (*Cannabis sativa* L.) in China." *Indian Journal of Traditional Knowledge* 16, no. 2 (2017): 235–42.

Liu, Guozhong 劉國忠. *Wuxing dayi yanjiu* 五行大義研究 (A Study of the General Principles of the Five Phases). Shenyang: Liaoning jiaoyu chubanshe, 1999.

——— "Shi lun shi'er shengxiao yu sanshiliu qin 試論十二生肖與三十六禽 (A Preliminary Study on Twelve Zodiac Animals and Thirty-Six Birds)." *Qinghua daxue xuebao* 清華大學學報 14, no. 1 (1999): 12–15.

Liu, William Guanglin 劉光臨. 2015. *The Chinese Market Economy, 1000–1500*. Albany: State University of New York Press.

Liu, Xianbo 劉顯波 and Xiong Jun 熊雋. *Tangdai jiaju yanjiu* 唐代傢俱研究 (*A Study on Tang Furniture*). Beijing: Renmin chubanshe, 2017.

Liu, Yan 劉焱. *Healing with Poisons: Potent Medicines in Medieval China*. Seattle: University of Washington Press, 2021.

Liu, Yonghua 劉永華. *Zhongguo gudai cheyu maju* 中國古代車輿馬具 (*Carts, Carriages, and Horse Saddles in Traditional China*). Shanghai: Shanghai cishu chubanshe, 2002.

Liu, Zhaomin 劉昭民 ed. *Zhonghua qixiang xue shi* 中華氣象學史 (*The History of Chinese Meteorology*). Taipei: Shangwu yinshu guan, 1970.

Liu, Zhaomin 劉昭民 and Liu Tangrui 劉棠瑞. *Zhonghua shengwuxue shi* 中華生物學史 (*A History of Chinese Biology*). Taipei: Taiwan shangwu yinshuguan 1991.

Liu, Zhongyu 劉仲宇. *Zhongguo jingguai wenhua* 中國精怪文化 (*Chinese Spirit and Anomaly Culture*). Shanghai: Shanghai renmin chubanshe, 1997.

Lobban Jr., Richard A. "Pigs and Their Prohibition." *International Journal of Middle East Studies* 26, no. 1 (1994): 57–75.

Loewe, Michael. *Dong Zhongshu 董仲舒, a "Confucian" Heritage and the Chunqiu Fanlu*. Leiden: Brill, 2011.

——— *Divination, Mythology and Monarchy in Han China*. Cambridge: Cambridge University Press, 1994.

Long, Sheng 龍聖 and Li Xiangzhen 李向振. "Binghuan: Bianpo gushi de shehui yinyu 病患:變婆故事的社會隱喻 (The Illed Patients: The Social Metaphor of the Tales on the Human Transformations)." *Minzu wenxue yanjiu* 民族文學研究 (*Ethnical Literature Studies*) 37, no. 3 (2019): 5–15.

Lü, Zongli 吕宗力. *Power of the Words: Chen Prophecy in Chinese Politics, AD 265–618*. Bern and New York: Peter Lang, 2003.

Luo, Guihuan 羅桂環. "Biology." In *A History of Chinese Science and Technology*, vol. 1, edited by Lu Yongxiang, 431–95. Shanghai: Shanghai Jiaotong University Press, Berlin and Heidelberg: Springer Verlag, 2015.

——— "Biology in Song China." In *Historical Perspectives on East Asian Science, Technology and Medicine*, edited by Chan Alan Kam Leung, Clancey Gregory K, and Loy Hui-chieh, 171–81. Singapore: World Scientific, 2002.

Lutwyche, Richard. *The Pig: A Natural History*. Princeton: Princeton University Press, 2019.

Lynteris, Christos ed. *Framing Animals as Epidemic Villains: Histories of Non-Human Disease Vectors*. Cham: Palgrave Macmillan, 2019.

Mair, Victor H. and Erling Hoh. *The True History of Tea*. New York and London: Thames & Hudson, 2009.

Mair, Victor H., Nancy S. Steinhardt, and Paul R. Goldin eds. *Hawai'i Reader in Traditional Chinese Culture*. Honolulu: University of Hawai'i Press, 2005.

Major, John S. *Heaven and Earth in Early Han Thought: Chapters Three, Four, and Five of the Huainanzi*. Albany: State University of New York Press, 1993.

Mann, Michael E. et al. "Proxy-Based Reconstructions of Hemispheric and Global Surface Temperature Variations over the Past Two Millennia." *Proceedings of the National Academy of Sciences* 105, no. 36 (2008): 13252–57.

Marks, Robert B. *China: Its Environment and History*. Lanham: Lexington Books, 2011.

——— *Tigers, Rice, Silk, and Silt: Environment and Economy in Late Imperial South China*. Cambridge: Cambridge University Press, 1998.

Martin, Dan. "Tibet at the Center – A Historical Study of Some Tibetan Geographical Conceptions Based on Two Types of Country-Lists Found in Bon Histories." In *Tibetan Studies: Proceedings of the 6th Seminar of the International Association for Tibetan Studies*, vol. 1, edited by Per Kværne, 517–32. Oslo: The Institute for Comparative Research in Human Culture.

McDermott, Joseph P. *A Social History of the Chinese Book: Books and Literati Culture in Late Imperial China*. Hong Kong: University of Hong Kong Press, 2006.

Min, Xiangpeng 閔祥鵬. "Gudai qixiang guance, minjian Xinyang yu wenxue chuangzuo de duoxueke shijiao: Xiyouji Zhong Zhubajie shuishen xingxiang de yanbian 古代氣象觀測,民間信仰與文學創作的多學科視角,西遊記中豬八戒水神形象的演變 (Multi-Disciplinary Perspectives of Ancient Climate Observations, Folk Beliefs, and Literary Creative Writings: The Changing Images of Zhu Bajie as a Water God in the *Journey to the West*). *Beijing shehui kexue* 北京社會科學, no. 10 (2014): 67–72.

——— "Cong Pei Xingjian yuce baoyu tanqi 從裴行儉預測暴雨談起 (On Pei Xingjian Forecasting a Rainstorm)." *Zhongguo jianzai* 中國減災 (*Natural Disaster Reduction in China*), no. 8 (2009): 43–44.

Min, Xiangpeng 閔祥鵬 and Huaiyu Chen. "Animals, Divination, and Climate: An Environmental Note on the Cult of the Pig in Ancient China." *ISLE: Interdisciplinary Studies in Literature and Environment* 28 (2021), https://doi.org/10.1093/isle/isaa195.

Minakata, Kumagusu 南方熊楠. *Minakata Kumagusu zenshū* 南方熊楠全集. Tokyo: Heibonsha, 1971.

Mizelle, Bret. *Pig*. London: Reaktion Books, 2011.

Mizuguchi, Motoki 水口幹記. *Nihon Kodai Kaiseki juyō no shiteki kenkyū* 日本古代漢籍受容の史的研究 (A Study on the History of the Acceptance of Classical Chinese Texts in Ancient Japan). Tōkyō: Kyūko Shoin, 2005.

Moberg, Anders et al. "Highly Variable Northern Hemisphere Temperatures Reconstructed from Low and High Resolution Proxy Data." *Nature*, no. 433 (2005): 613–17.

Nakamura, Zuiryu 中村瑞隆, "Hōshō daranikyō dankan 寶星陀羅尼經斷簡 (The Fragment of the Ratnaketudhāraṇī)." *Hokekyô Bunka Kenkyû*法華文化研究, no. 1 (1975): 13–45.

Nattier, Jan. *A Guide to the Earliest Chinese Buddhist Translations: Texts from the Eastern Han and Three Kingdoms Periods.* Bibliotheca Philologica et Philosophica Buddhica X, Tokyo: The International Research Institute for Advanced Buddhology, Soka University, 2008.

——— *Once Upon a Future Time: Studies in a Buddhist Prophecy of Decline.* Berkeley: Asian Humanities Press, 1991.

Nawa, Toshimitsu 名和敏光 ed. *Higashi Ajia shisō – Bunka no kisō kōzō: jussū to Tenchi zuishōshi* 東アジア思想•文化の基層構造:術数と『天地瑞祥志』 [The Foundational Structure of East Asian Thoughts and Culture: Numbers and Techniques and the *Treatise on the Auspicious Signs of Heaven and Earth*]. Tōkyō: Kyūkoshoin, 2019.

Needham, Joseph. *Science and Civilisation in China. Volume 6: Biology and Biological Technology. Part I: Botany.* With the Collaboration of Lu Gwei-Djen and a special contribution by Huang Hsing-tsung. Cambridge: Cambridge University Press, 1996.

Newman, Patrick. *Tracking the Weretiger: Supernatural Man-Eaters of India, China and Southeast Asia.* Jefferson, NC and London: McFarland & Co., 2012.

Nicol, Janine. "Outsiders: Medieval Chinese Buddhists and the 'Borderland Complex' – An Exploration of the Eight Difficulties," *The SOAS Journal of Postgraduate Research* 6 (2014): 27–48.

Nylan, Michael. "Humans as Animals and Things in Pre-Buddhist China." *Religions* 10 (2019): 360; doi:10.3390/rel10060360.

——— "The 'Treatise on the Wuxing' (Wuxingzhi 五行誌)." In *Monographs in Tang Official Historiography: Perspectives from the Technical Treatises of the History of Sui* (*Sui shu* 隋書), edited by Daniel Patrick Morgan, Damien Chaussende, and Karine Chamla, 181–236. Dordrecht: Springer Nature, 2019.

Ohnuma, Reiko. *Unfortunate Destiny: Animals in the Indian Buddhist Imagination.* New York: Oxford University Press, 2017.

Peterson, Anna. "Religious Studies and the Animal Turn." *History of Religions* 56, no. 2 (2016): 232–45.

Poo, Mu-chou 蒲慕洲. *Ghosts and Religious Life in Early China.* Cambridge: Cambridge University Press, 2022.

——— ed., *Rethinking Ghosts in World Religions.* Leiden: Brill, 2009.

——— "Zhongguo gudai gui lunshu de xingcheng 中國古代鬼論述的形成 (The Formation of the Concept of Ghost in Ancient China)." In *Guimeishenmo: Zhongguo tongsu wenhua cexie* 鬼魅神魔:中國通俗文化側寫 (Ghost, Enchantment, God, Demon: Perspectives on the Chinese Popular Culture), edited by Poo Mu-chou, 19–40. Taipei: Maitian chubanshe, 2005.

——— "The Completion of an Ideal World: The Human Ghost in Early Medieval China." *Asia Major* 3rd series, 10 (1997): 69–94.

——— "Ghost Literature: Exorcistic Ritual Texts or Daily Entertainment?" *Asia Major* 3rd series 13, no. 1 (2000): 43–64.

Puett, Michael. *To Become a God: Cosmology, Sacrifice, and Self-Divinization in Early China.* Cambridge: Harvard University Asia Center, 2002.

Raphals, Lisa. *Divination and Prediction in Early China and Ancient Greece.* Cambridge: Cambridge University Press, 2013.

Rong, Xinjiang 榮新江. 2013. *Eighteen Lectures on Dunhuang.* Translated by Imre Galambos. Leiden: Brill.

Sadakata, Akira 定方晟. *Buddhist Cosmology: Philosophy and Origins.* Translated by Gaynor Sekimori, with a foreword by Hajime Nakamura. Tokyo: Kōsei Publishing Co., 1997.

Sagart, Laurent. "The Chinese Names of the Four Directions." *Journal of American Oriental Society* 124, no. 1 (2004): 69–76.

Sawada, Mizuno 澤田瑞穗. *Chūgoku dōbutsutan* 中國動物譚 (*Chinese Animal Tales*). Tokyo: Kōbundō, 1978.

Schaefer, Donovan. *Religious Affects: Animality, Evolution, and Power.* Durham: Duke University Press, 2015.

Schafer, Edward H. *The Golden Peaches of Samarkand: A Study of T'ang Exotics.* Berkeley: University of California Press, 1963.

Schmidt, Gerhard and Thomas Thilo. *Katalog chinesischer buddhistischer Textfragmente I* (BTT VI). Berlin: Wiley-VCH Verlag, 1971.

Schmitt, Jean-Claude. *Ghosts in the Middle Ages: The Living and the Dead in Medieval Society.* Chicago: University of Chicago Press, 1998.

Schneider, Mindi. "Reforming the Humble Pig: Pigs, Pork, and Contemporary China." In *Animals through Chinese History: Earliest Times to 1911*, edited by Roel Sterckx, Martina Siebert, Dagmar Schäfer, 233–43. Cambridge: Cambridge University Press, 2018.

Shahar, Meir. "The Chinese Cult of the Horse King, Divine Protector of Equines." In *Animals and Human Society in Asia: Historical, Cultural and Ethical Perspectives*, edited by Rotem Kowner, Guy Bar-Oz, Michal Biran, Meir Shahar, and Gideon Shelach, 355–90. Dordrecht: Springer, 2019.

Shao, Xiaofeng 邵曉峰. *Zhongguo Songdai jiaju: Yanjiu yu tuxiang jicheng* 中國宋代傢俱:研究與圖像集成 (*Furnitures in Song China: Studies and Collected Images*). Nanjing: Dongnan daxue chubanshe, 2010.

Shaughnessy, Edward L. *Unearthing the Changes: Recently Discovered Manuscripts of the Yi Jing (I Ching) and Related Texts.* New York: Columbia University Press, 2014.

Shen, Hsueh-man 沈雪曼. *Gilded Splendor: Treasures of China's Liao Empire (907–1125).* Milano: 5 Continents, 2006.

Sheng, Angela 盛餘韻. "Determining the Value of Textiles in the Tang Dynasty: In Memory of Professor Denis Twitchett (1925–2006)." *Journal of the Royal Asiatic Society* 3rd series 23, no. 2 (2013): 175–95.

Shimamori, Tetsuo 島森哲男. "Chugoku no hito tora henshintan: 'nugu/Kiru' no minwagaku 中国の人虎変身譚ー「脱ぐ / 着る」の民話学 (Chinese Tales of the Tiger-Human Transformations: Wearing/Taking off in the Folklore)." *Miyagi Kyōiku Daigaku kiyō* 宮城教育大学紀要, no. 53 (2018), 403–26.

Siklos, Bulcsu. "The Evolution of the Budhist Yama." In *Buddhist Forum,* vol. IV, edited by Tadeusz Skorupski, 165–89. London: School of Oriental and African Studies, University of London, 1996.

Simmons, Frederick J. "Traditional Use and Avoidance of Foods of Animal Origin: A Culture Historical View." *BioScience* 28, no. 3 (1978): 178–84.

Singh, Ananda. *Planet, Plants and Animals: Ecological Paradigms in Buddhism.* Delhi: Primus Books, 2019.

Smith, Richard J. *Fortune-Tellers and Philosophers: Divination in Traditional Chinese Society.* Boulder: Westview Press, 1991.

Snyder-Reinke, Jeffrey P. *Dry Spells: Rainmaking, Power, and the State in Late Imperial China.* Cambridge, MA: Harvard University Asia Center, 2006.

Songster, E. Elena. *Panda Nation: The Construction and Conservation of China's Modern Icon.* New York: Oxford University Press, 2018.

Snowden, Frank M. *Epidemics and Society: From the Black Death to the Present.* New Haven: Yale University Press, 2019.

Stein, Rolf A. *The World in Miniature: Container Gardens and Dwellings in Far Eastern Religious Thought*, translated by Phyllis Brooks with a New Foreword by Edward H. Schafer. Stanford: Stanford University Press, 1990.

Steinhardt, Nancy Shatzman. "The Tang Architectural Icon and the Politics of Chinese Architectural History." *The Art Bulletin* 86, no. 2 (2004): 228–54.

Sterckx, Roel. *The Animals and the Daemon in Early China.* State University of New York Press, 2002.

——— "Animal Classification in Ancient China." *East Asian Science, Technology and Medicine* 23 (2005): 96–123.

Sterckx, Roel, Martina Siebert, Dagmar Schäfer eds. *Animals through Chinese History: Earliest Times to 1911.* Cambridge: Cambridge University Press, 2018.

Straughan, Roger. *Ethics, Morality and Animal Biotechnology.* Swindon: Biotechnology and Biological Sciences Research Council, 1999.

Strickmann, Michel. *Mantras et mandarins: le bouddhisme tantrique en Chine.* Paris: Gallimard, 1996.

——— "The *Consecration Sūtra*: A Buddhist Book of Spells." In *Chinese Buddhist Apocrypha*, edited by Robert E. Buswell Jr. and Kyoko Tokuno, 75–118. Honolulu: University of Hawai'i Press, 1990.

Suzuki, Yūiki 鈴木陽一. "Seiūki niokeru jinbutsu keishū no seikentō 2 Chūhakka to Songoku 西遊記における人物形象の再檢討 (2) 豬八戒と孫悟空 (A Reexamination of the Images of Figures in the Journey to the West 2: Zhu Bajie and Sun Wukong)." *Gengo bunka kenkyū* 言語文化研究 (*Studies on Languages and Cultures*) 1, no. 1 (1981): 123–50.

Teiser, Stephen F. *Reinventing the Wheel: Paintings of Rebirth in Medieval Buddhist Temples.* Seattle and London: University of Washington Press, 2007.

——— *The Ghost Festival in Medieval China.* Princeton: Princeton University Press, 1988.

Thomas, Craig and David Pijawka. "Studying American Naturalists for Transdisciplinary Solutions to Today's Wicked Problems." *ISLE: Interdisciplinary Studies in Literature and Environment* 27, no. 2 (Spring 2020): 221–42.

Tokura, Hidemi 戶倉英美. "Bianshen gushi de bianqian: you liuchao zhiguai xiaoshuo dao Liaozhai zhiyi 變身故事的變遷：由六朝志怪小說到聊齋誌異 (The Evolution of the Tales of Body Transformation: From the Anomaly Tale in the Six Dynasties to the Strange Tales in a Chinese Studio)." In *Guoji liaozhai lunwen ji* 國際聊齋論文集 (*Collected Papers from the International Studies on Strange Tales from a Chinese Studio*), edited by Gu Meigao 辜美高 and Wang Zhizhong 王枝忠, 161–98. Beijing: Beijing shifan xueyuan chubanshe, 1992.

Tominaga, Kazuto 富永一登. "Jin toraten no keifu: Rikuchō katoraten kara Tōten shōsetsue 人虎傳の系譜—六朝化虎譚から唐傳小說へ (The Geaology of the Weretiger Tales: From the Tales of the Tiger Transformation to the Stories in Tang Legends)." *Chūgoku chūsei bungaku kenkyū* 中國中世文學研究 13 (1978): 1–17.

Tripāṭhī, Chandrabhāl. *Ekottarāgama-Fragmente der Gilgit-Handschrift.* Studien zur Indologie und Iranistik, Monographien 2. Reinbek: Dr. Inge Wezler, Verlag für Orientalische Fachpublikationen, 1995.

Trombert, Eric. "Cooking, Dyeing, and Worship: The Uses of Safflower in Medieval China as Reflected in Dunhuang Documents." *Asia Major* 3rd series 17, no. 1 (2004): 59–72.

——— "Des fleurs rouges en galette: Une plante tinctoriale dans la Chine ancienne: le carthame." *Journal Asiatique* 285, no. 2 (1997): 509–47.

Tsinghua Daxue jianzhu sheji yanjiuyuan 清華大學建築設計研究院 (Institute of Architectural Design at Tsinghua University), and Beijing Qinghua chengshi guihua sheji yanjiuyuan 北京清華城市規劃設計研究院 (Institute of Urban Planning and Design at Tsinghua, Beijing). *Foguangsi dongdadian jianzhu kancha yanjiu* 佛光寺東大殿建築勘查研究 (*An Examination and Study of the East Hall of Buddha's Light Temple*). Beijing: Wenwu chubanshe, 2011.

Tu, Cheng-sheng 杜正勝. "Gudai wuguai zhi yanjiu: yizhong xintaishi he wenhuashi de tansuo 古代物怪之研究：一種心態史和文化史的探索 (A Study on Object Spirits in Ancient China: An Investigation from the Perspective of the History of Mentality and Cultural History)." (Three parts). *Dalu zazhi* 大陸雜誌 (*Mainland Journal*) 104, no. 1 (2001): 1–14; 104, no. 2 (2001): 1–15; 104, no. 3 (2001): 1–10.

Tudkeao, Chanwit. "Three Sanskrit Fragments of the *Ratnaketuparivarta*." In *The British Library Sanskrit Fragments*, vol. 3, edited by Seishi Karashima and Klaus Wille, 589–91. Tokyo: International Research Institute for Advanced Buddhology, Soka University, 2015.

Twitchett, Denis. *Printing and Publishing in Medieval China*. New York: Frederic C. Bell, 1983.

van Ess, Hans. "The Apocryphal Texts of the Han Dynasty and the Old Text/New Text Controversy." *T'oung Pao* 85, nos. 1–3 (1999): 29–64.

Verellen, Franciscus. *Imperiled Destinies: The Daoist Quest for Deliverance in Medieval China*. Cambridge, MA: Harvard University Asia Center, 2019.

Vira, Raghu and Lokesh Chandra. *Gilgit Buddhist Manuscripts*. New Delhi: International Academy of Indian Culture, 1974.

von Glahn, Richard. *The Economic History of China: From Antiquity to the Nineteenth Century*. Cambridge: Cambridge University Press, 2006.

Walters, Terrence W. "Historical Overview on Domesticated Plants in China with Special Emphasis on the Cucurbitaceae." *Economic Botany* 43, no. 3 (1989): 297–313.

Wang, Aihe 王愛和. *Cosmology and Political Culture in Early China*. Cambridge: Cambridge University Press, 2000.

Wang, Binghua 王炳華 and Helen Wang. "A Study of the Tang Dynasty Tax Textiles (Yongdiao Bu) from Turfan." *Journal of the Royal Asiatic Society* 3rd series 23, no. 2 (2013): 263–80.

Wang, Fengling 汪玢玲. Zhongguo huwenhua 中國虎文化 (*Tiger Culture in China*). Beijing: Zhonghua shuju, 2007.

Wang, Hongjun 王洪軍. "Tangdai de chaye shengchan 唐代的茶葉生產 (The Tea Production in the Tang Dynasty)." *Qilu xuekan* 齊魯學刊 (*Journal of Qi Lu*), no. 6 (1987): 14–21.

Wang, Jiakui 王家葵 and Zhang Ruixian 張瑞賢. *Shennong bencaojing yanjiu* 神農本草經研究 (A Study of *Herbal Classic of the Divine Farmer*). Beijing: Beijing kexue jishu chubanshe, 2001.

Wang, Li 王立. "Liaozhai zhiyi Huagao huahu fuchou gushi de Zhong Yin wenxue yuanyua n聊齋誌異向杲化虎復仇故事的中印文學淵源 (The Sino-Indian Literary Sources for the Tale of the Tiger Transformation of Xiang Gao for Taking Revenge in the Strange Tales from a Chinese Studio)." *Huanan shifan daxue xuebao* 華南師範大學學報, no. 1 (2008): 59–66.

Waley, Arthur trans. *Book of Songs*. London: Routledge, 2005, reprinted edition.

Wayman, Alex. "Studies in Yama and Māra." *Indo-Iranian Journal* 3 (1959): 44–73.

Wen, Yiduo 聞一多. Wen *Yiduo quanji* 聞一多全集 (*The Collected Works of Wen Yiduo*). Beijing: Sanlian shudian, 1982.

Wessing, Robert. "Homo Narrans in East Java: Regional Myths and Local Concerns." *Asian Folklore Studies* 65, no. 1 (2006): 45–68.

——— "The Last Tiger in East Java: Symbolic Continuity in Ecological Change." *Asian Folklore Studies* 54 (1995): 191–218.

——— "'Bangatowa,' 'Patogu' and 'Gaddhungan': Perceptions of the Tiger among the Madurese." *Journal of Southeast Asian Studies* 25, no. 2 (1994): 368–80.

Williams, David. *Deformed Discourse: The Function of the Monster in Mediaeval Thought and Literature.* Montreal: McGill-Queen's University Press, 1999.

Wilson, J. Keith. "Powerful Form and Potent Symbol: The Dragon in Asia." *The Bulletin of the Cleveland Museum of Art* 77, no. 8 (1990): 286–323.

Winzeler, Robert L. *Anthropology and Religion: What We Know, Think, and Question*, 2nd edition. Lanham: Altamira Press, 2012.

Wolf, Arthur P. "Gods, Ghosts, and Ancestors." In *Religion and Ritual in Chinese Society*, edited by Arthur P. Wolf, 131–82. Stanford: Stanford University Press, 1974.

Woods, Abigail, Michael Bresalier, Angela Cassidy, and Rachel Mason Dentinger eds. *Animals and the Shaping of Modern Medicine: One Health and Its Histories.* London: Palgrave Macmillan, 2018.

Xin, Yu 余欣. "Liuli in Buddhist Rituals and Art in Medieval China." *Hualin International Journal of Buddhist Studies* 1, no. 2 (2018): 231–68.

——— *Zhonggu yixiang: Xieben shidai de xueshu, Xinyang yu Shehui* 中古異相：寫本時代的學術、信仰與社會 (*Mapping the Extraordinary: Knowledge, Faith and Manuscript Culture in Medieval China*). Shanghai: Shanghai guji chubanshe, 2011.

Yan, Yaozhong 嚴耀中. "Lun zhanbu yu Sui Tang fojiao de jiehe 論占卜與隋唐佛教的結合 [On the Combination between Divination and Buddhism in the Sui and Tang Dynasties]. *Shijie zongjiao yanjiu* 世界宗教研究 4 (2002): 30–37.

Yancheva, Gergana et al. "Influence of the Intertropical Convergence Zone on the East Asian Monsoon." *Nature* 445, no. 7123 (2007): 74–77.

Yang, Bao et al. "General Characteristics of Temperature Variation in China during the Last Two Millennia." *Geophysical Research Letters* 29, no. 9 (2002): 38(1–4).

Yang, Gang and Christoph Anderl. "Prognostication in Chinese Buddhist Historical Texts: The *Gaoseng zhuan* and the *Xu Gaoseng zhuan*." *Acta Orientalia Academiae Scientiarum Hungaricae* 73, no. 1 (2020): 1–45.

Yang, Xiaoshan 楊曉山. "Li Gefei's 'Luoyang mingyuan ji 洛陽名園記' (A Record of the Celebrated Gardens of Luoyang) Text and Context." *Monumenta Serica* 52 (2004): 221–55.

——— "Li Deyu's 李德裕 Pingquan Villa: Forming an Emblem from the Tang and to the Song." *Asia Major* 3rd series 17, no. 2 (2004): 45–88.

Yao, Lijiang 姚立江, Pan Lanxiang 潘蘭香. *Renwen dongwu: Dongwu fuhao yu zhongguo wenhua* 人文動物：動物符號與中國文化 (*Humanistic Animals: The Animal Signs and Chinese Culture*). Haerbin: Heilongjiang renmin chubanshe, 2002.

Yü, Chün-fang 于君方. *Chinese Buddhism: A Thematic History.* Honolulu: University of Hawai'i Press, 2020.

Zeng, Xiongsheng 曾雄生. "Agriculture." In *A History of Chinese Science and Technology*, vol. 1, edited by Lu Yongxiang, 351–429. Shanghai: Shanghai Jiaotong University Press; Berlin and Heidelberg: Springer Verlag, 2015.

Zhang, Dexin 張德鑫. "Fangweici de wenhua kaocha 方位詞的文化考察 (A Cultural Perspective on the Vocabulary of Cardinal Directions)." *Shijie hanyu jiaoxue* 世界漢語教學 37, no. 3 (1996): 62–73.

Zhang, Ling 張玲. *The River, the Plain, and the State: An Environmental Drama in Northern Song China*. Cambridge: Cambridge University Press, 2016.

Zhang, Xing 張幸 and Chen Huaiyu. "From Lion to Tiger: The Buddhist Changing Images of Apex Predators in Trans-Asian Contexts." In *Animals and Human Society in Asia: Historical and Ethical Perspectives*, edited by Rotem Rosen, Michal Biran, Meir Shahar, and Gideon Shelach, 331–53. Basingstoke: Palgrave Macmillan, 2019.

Zhang, Yun 張蘊. "Xi'an diqu Sui Tang muzhi wenshi zhong de shi'er shengxiao tu'an 西安地區隋唐墓誌紋飾中的十二生肖圖案 (The Twelve-Zodiac Animal Pictures in the Patterns Decorated the Tombstones in the Sui and Tang Tombs in the Xi'an Region)." *Tang Yanjiu* 唐研究 8 (2002): 395–432.

Zhang, Zehong 張澤洪. "Zhongguo xinan shaoshu minzu zongjiao zhong de huchongbai yanjiu 中國西南少數民族宗教中的虎崇拜研究 (A Study on the Tiger Cult among the Minority Religions in Southwest China." *Zhongnan minzu daxue xuebao* 中南民族大學學報, no. 6 (2007): 38–43.

Zhao, Dandan, Jie Wang, Yanjing, Cui, and Xinfang Wu. "Pharmacological Effects of Chinese Herb Aconite (Fuzi) on Cardiovascular System." *Journal of Traditional Chinese Medicine* 32, no. 3 (2012): 308–13.

Zhao, Lu. *In Pursuit of the Great Peace: Han Dynasty Classicism and the Making of Early Medieval Literati Culture*. Albany: State University of New York Press, 2019.

Zhao, Xianhuan. *Classical Chinese Supernatural Fiction: A Morphological History*. Lewiston: E. Mellen Press, 2005.

Zhong, Z. P., Tian, F., Roux, S. et al. "Glacier Ice Archives Nearly 15,000-Year-Old Microbes and Phages." *Microbiome* 9, no. 1 (2021): 160. https://doi.org/10.1186/s40168-021-01106-w.

Zhou, Xiaolu 周曉陸. "Shi Dongxinanbeizhong: jianshuo ziwu 釋東西南北中:兼說子午 (Notes on East, West, South, North and Center: Also on Zi Wu)." *Nanjing daxue xuebao* 南京大學學報, no. 3 (1996): 70–76.

Zhu, Kezhen 竺可楨. "Zhongguo jin wuqiannian lai qihou bianqian de chubu yanjiu 中國近五千年來氣候變遷的初步研究 (A Preliminary Study on the Climate Change in China in Nearly Five Thousand Years)." *Kaogu xuebao* 考古學報 (*Archaeological Journal*) 1 (1972): 15–38.

Zhu, Kezhen 竺可楨 and Wan Minwei 宛敏渭. *Wuhou xue* 物候學 (Phenology). Expanded edition 增訂本. Beijing: Kexue chubanshe, 1973.

Zinsstag, Jacob, Esther Schelling, David Waltner-Toews, Maxine Whittaker, and Marcel Tanner eds. *One Health: The Theory and Practice of Integrated Health Approaches*. Wallingford: CAB International, 2015.

Zurndorfer, Harriet T. "Cotton Textile Manufacture and Marketing in Late Imperial China and the 'Great Divergence'." *Journal of the Economic and Social History of the Orient* 54, no. 5 (2011): 701–38.

INDEX

www.ingramcontent.com/pod-product-compliance
Ingram Content Group UK Ltd.
Pitfield, Milton Keynes, MK11 3LW, UK
UKHW011005120926
8556IPUK00027B/116

9 781839 985010